Advances in Smart and Tough Hydrogels

Advances in Smart and Tough Hydrogels

Editors

Dong Zhang
Jintao Yang
Xiaoxia Le
Dianwen Song

Basel • Beijing • Wuhan • Barcelona • Belgrade • Novi Sad • Cluj • Manchester

Editors

Dong Zhang
Georgia Institute of
Technology and
Emory University
Atlanta
United States

Jintao Yang
College of Materials Science
& Engineering
Zhejiang University
of Technology
Hangzhou
China

Xiaoxia Le
Ningbo Institute of Material
Technology and Engineering
Chinese Academy of Sciences
Ningbo
China

Dianwen Song
School of Medicine
Shanghai Jiaotong University
Shanghai
China

Editorial Office
MDPI
St. Alban-Anlage 66
4052 Basel, Switzerland

This is a reprint of articles from the Special Issue published online in the open access journal *Gels* (ISSN 2310-2861) (available at: www.mdpi.com/journal/gels/special_issues/smart_tough_hydrogels).

For citation purposes, cite each article independently as indicated on the article page online and as indicated below:

Lastname, A.A.; Lastname, B.B. Article Title. *Journal Name* **Year**, *Volume Number*, Page Range.

ISBN 978-3-0365-9972-4 (Hbk)
ISBN 978-3-0365-9971-7 (PDF)
doi.org/10.3390/books978-3-0365-9971-7

© 2024 by the authors. Articles in this book are Open Access and distributed under the Creative Commons Attribution (CC BY) license. The book as a whole is distributed by MDPI under the terms and conditions of the Creative Commons Attribution-NonCommercial-NoDerivs (CC BY-NC-ND) license.

Contents

About the Editors vii

Dong Zhang, Jintao Yang, Xiaoxia Le and Dianwen Song
Editorial for Special Issue: Advances in Smart and Tough Hydrogels
Reprinted from: *Gels* **2023**, *9*, 789, doi:10.3390/gels9100789 1

Yutaka Ohsedo and Wakana Ueno
Creation of Polymer Hydrogelator/Poly(Vinyl Alcohol) Composite Molecular Hydrogel Materials
Reprinted from: *Gels* **2023**, *9*, 679, doi:10.3390/gels9090679 6

Muhammad Suhail, I-Hui Chiu, Yi-Ru Lai, Arshad Khan, Noorah Saleh Al-Sowayan and Hamid Ullah et al.
Xanthan-Gum/Pluronic-F-127-Based-Drug-Loaded Polymeric Hydrogels Synthesized by Free Radical Polymerization Technique for Management of Attention-Deficit/Hyperactivity Disorder
Reprinted from: *Gels* **2023**, *9*, 640, doi:10.3390/gels9080640 17

Romana Kratochvílová, Milan Kráčalík, Marcela Smilková, Petr Sedláček, Miloslav Pekař and Elke Bradt et al.
Functional Hydrogels for Agricultural Application
Reprinted from: *Gels* **2023**, *9*, 590, doi:10.3390/gels9070590 31

Marin Simeonov, Anton Atanasov Apostolov, Milena Georgieva, Dimitar Tzankov and Elena Vassileva
Poly(acrylic acid-co-acrylamide)/Polyacrylamide pIPNs/Magnetite Composite Hydrogels: Synthesis and Characterization
Reprinted from: *Gels* **2023**, *9*, 365, doi:10.3390/gels9050365 48

Yazhou Wang, Zhiwei Peng, Dong Zhang and Dianwen Song
Tough, Injectable Calcium Phosphate Cement Based Composite Hydrogels to Promote Osteogenesis
Reprinted from: *Gels* **2023**, *9*, 302, doi:10.3390/gels9040302 70

Bianca-Elena-Beatrice Crețu, Alina Gabriela Rusu, Alina Ghilan, Irina Rosca, Loredana Elena Nita and Aurica P. Chiriac
Cryogel System Based on Poly(vinyl alcohol)/Poly(ethylene brassylate-co-squaric acid) Platform with Dual Bioactive Activity
Reprinted from: *Gels* **2023**, *9*, 174, doi:10.3390/gels9030174 82

Hideaki Tokuyama, Ryo Iriki and Makino Kubota
Thermosensitive Shape-Memory Poly(stearyl acrylate-*co*-methoxy poly(ethylene glycol) acrylate) Hydrogels
Reprinted from: *Gels* **2023**, *9*, 54, doi:10.3390/gels9010054 94

Tim B. Mrohs and Oliver Weichold
Hydrolytic Stability of Crosslinked, Highly Alkaline Diallyldimethylammonium Hydroxide Hydrogels
Reprinted from: *Gels* **2022**, *8*, 669, doi:10.3390/gels8100669 101

Bailin Dai, Ting Cui, Yue Xu, Shaoji Wu, Youwei Li and Wu Wang et al.
Smart Antifreeze Hydrogels with Abundant Hydrogen Bonding for Conductive Flexible Sensors
Reprinted from: *Gels* **2022**, *8*, 374, doi:10.3390/gels8060374 . . . . **112**

Tianzhu Shi, Zhengfeng Xie, Xinliang Mo, Yulong Feng, Tao Peng and Dandan Song
Highly Efficient Adsorption of Heavy Metals and Cationic Dyes by Smart Functionalized Sodium Alginate Hydrogels
Reprinted from: *Gels* **2022**, *8*, 343, doi:10.3390/gels8060343 . . . . **123**

About the Editors

Dong Zhang

Dong Zhang received his Ph.D. degree in chemical engineering from the University of Akron (with Professor Jie Zheng) in December 2022. He has been working as a postdoctoral fellow at Professor Younan Xia's group in the Georgia Institute of Technology and Emory University since January 2023. His research interests include the rational design and synthesis of polymers, biomaterials, hydrogels, and nanostructured materials for environmental, biological, and biomedical applications. He has co-authored more than 100 peer-reviewed publications.

Jintao Yang

Dr. Jintao Yang is a Professor within the College of Materials Science and Technology, Zhejiang University of Technology. He received his B.S. and Ph.D. degrees in chemical engineering at China University of Petroleum (2000) and Zhejiang University (2005), respectively. His main research interests include polymer processing, polymer surfaces, and interfaces, particularly smart materials based on zwitterionic polymers for biological and sensing applications.

Xiaoxia Le

Dr. Xiaoxia Le received her Ph.D. degree in polymer chemistry and physics from the Ningbo Institute of Materials Technology and Engineering (NIMTE), Chinese Academy of Sciences (2019). She then joined Professor Tao Chen's group as a postdoctoral research fellow. Currently, she is an associate professor at NIMTE. Her research focuses on the construction and functionalization of stimuli-responsive hydrogels for applications in the soft actuator and fluorescent information anti-counterfeiting fields.

Dianwen Song

Dr. Dianwen Song is a Professor within the School of Medicine, Shanghai Jiaotong University. His research interests include bone tissue engineering, bio-mimetic hydrogel, and 3D printing. He has published over 60 peer-reviewed papers.

Editorial

Editorial for Special Issue: Advances in Smart and Tough Hydrogels

Dong Zhang [1,*], Jintao Yang [2], Xiaoxia Le [3] and Dianwen Song [4]

1. The Wallace H. Coulter Department of Biomedical Engineering, Georgia Institute of Technology and Emory University, Atlanta, GA 30332, USA
2. College of Materials Science & Engineering, Zhejiang University of Technology, Hangzhou 310014, China; yangjt@zjut.edu.cn
3. Ningbo Institute of Materials Technology and Engineering, Chinese Academy of Sciences, Ningbo 315201, China; lexiaoxia@nimte.ac.cn
4. School of Medicine, Shanghai Jiaotong University, Shanghai 200240, China; dwsong@sjtu.edu.cn
* Correspondence: dzhang470@gatech.edu

1. Introduction

Smart hydrogels possess both intelligent and responsive properties, which are designed to exhibit specific responses to external stimuli such as changes in temperature, pH, or the presence of specific ions/counterions, making them "smart" or "responsive" materials. Tough hydrogels are engineered to have exceptional mechanical strength/toughness, which means they can endure significant mechanical stress or deformation without breaking or losing their structural integrity. Both of them have a wide range of potential applications in fields such as biomedicine, tissue engineering, drug delivery, and soft robotics, where their combination of responsiveness and mechanical resilience can be highly advantageous.

This Special Issue features contributions from prominent experts in the field, with substantial input from 53 researchers representing more than 10 diverse regions worldwide (Figure 1a). These regions encompass countries such as China (including Taiwan), Japan, the USA, Germany, Romania, Bulgaria, the Czech Republic, Austria, Pakistan, and Saudi Arabia. Significantly, the important research collection included essential synthesis and characterization processes, along with practical applications spanning biological and tissue engineering, heavy metal removal, flexible sensors, shape memory devices, and agricultural fertilizers, among others (Figure 1b).

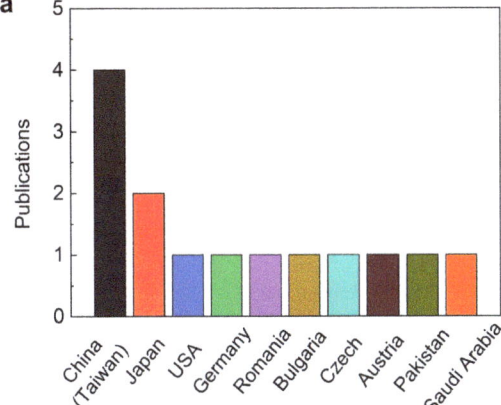

Figure 1. *Cont.*

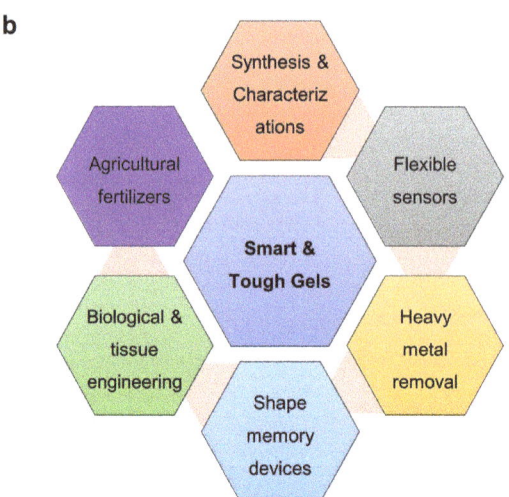

Figure 1. Summary of the *Advances in Smart and Tough Hydrogels* as a Special Issue of *Gels*. (**a**) Prominent contributing countries in research articles and (**b**) key themes in smart and tough hydrogels explored for this Special Issue.

2. Contributions

It is evident that over an extended period, the quest for a comprehensive understanding of gelation phenomena, coupled with the discovery of innovative gelation mechanisms and systems, has remained a pressing and unmet need. This ongoing pursuit underscores the continuous demand for advancements in this field, driven by the ever-evolving challenges and opportunities it presents. With this mission, Yutaka et al. contributed an intriguing gelation phenomenon: the intelligent gelation occurring at low concentrations when combining the polymer hydrogelator NaPPDT with the water-soluble polymer poly(vinyl alcohol) (PVA) [1]. This unique behavior is exclusive to the NaPPDT and PVA combination, as it is not observed when mixing aqueous anionic polymer solutions with other sodium sulfonate or phosphonic acid side chains and aqueous PVA solutions. More interestingly, even when diluting the concentrations of the gelator and PVA in aqueous solutions, the composite gel materials exhibit improved mechanical properties.

Not all chemical gelation processes exhibit stability. Indeed, the potential for degradable dynamics holds great promise, particularly in the context of injectable hydrogels. The Weichold group conducted a comprehensive assessment of the durability of alkaline hydrogels utilizing a widely used crosslinker, N,N′-methylenebisacrylamide (MBAA), and three recently introduced tetraallyl crosslinkers [2]. Upon subjecting the hydrogel models to accelerated aging at 60 °C for 28 days, those crosslinked with MBAA eventually transitioned into a liquid state, whereas the storage modulus and degree of swelling of the hydrogels crosslinked with the tetraallyl compounds which remained unaltered. This work significantly addresses the current knowledge gap in our understanding of the long-term performance and durability of synthetic hydrogels when subjected to hydrolytic conditions, while it simultaneously provides valuable insights into the practical utility and longevity of these crosslinked systems. Structurally, Marin et al. developed a novel hydrogel based on poly(acrylic acid-*co*-acrylamide)/polyacrylamide pseudo-interpenetrating polymer networks (pIPNs), with the inclusion of magnetite achieved through the in situ precipitation of Fe(III)/Fe(II) ions within the hydrogel matrix [3]. This unique design conferred upon the hybrid hydrogels the ability to respond to pH and ionic strength variations, in addition to endowing them with superparamagnetic properties. This breakthrough highlights the potential of pIPNs as matrices for precisely controlling the deposition of inorganic particles, representing a promising approach for manufacturing structural soft matter. In essence, researchers are encouraged to utilize chemical tools/reactions for fine-tuning network distributions and structures in the development of smart and robust hydrogels, such as phase separation or macrophase separation strategies [4–6].

In the fields of smart medicines and biomedical tissue engineering, Wu and coworkers engineered a smart hydrogel by melding xanthan gum, Pluronic F-127, and synthetic monomer acrylic acid using the free radical polymerization technique [7]. Smart hydrogels exhibited minimal swelling behavior at pH 1.2 and 4.6, while they displayed increased drug release at pH 7.4, indicating their pH-responsive characteristics in regulating the release of atomoxetine HCl within the colon over an extended duration. The resulting hydrogels demonstrated an impressive level of intelligence and adaptability, underscoring their potential for diverse applications in drug delivery and control release. Zhang, Song, and colleagues embarked on the mission to tackle existing challenges in local fracture healing and early anti-osteoporosis therapy for osteoporosis [8]. Their approach involved designing injectable hydrogels loaded with calcium phosphate cement (CPC). The integration of this robust biomimetic hydrogel with bioactive CPC represents a highly promising and innovative contender for the development of commercial clinical materials aimed at improving the prognosis of patients grappling with osteoporotic fractures.

As for the applications of environmental science, Shi et al. demonstrated the remarkable adsorption capabilities of smart hydrogels using a functionalized sodium alginate hydrogel (FSAH) [9]. This hydrogel was specifically engineered for the efficient removal of heavy metals and dyes by incorporating hydrazide-functionalized sodium alginate with hydrazone groups for the selective capture of heavy metals (Pb^{2+}, Cd^{2+}, and Cu^{2+}), while another dopamine grafting functional group offers active sites for adsorbing methylene blue, malachite green, and crystal violet. Notably, even after undergoing five adsorption–desorption cycles, it retained over 70% efficiency in the removal process. The encapsulation of hydrophobic molecular compounds into a polymer matrix has emerged as a method to modulate low solubility in water and a promising approach to preserve their chemical integrity, efficacy, but also their controlled release in a pulsating or continuous regime. A new cryogel system based on PVA and poly(ethylene brassylate-co-squaric acid) (PEBSA) obtained by repeated freeze–thaw processes, showing the cumulative antioxidant efficiency and antimicrobial activity against E. coli (Gram-negative strain), S. aureus (Gram-positive strain), and C. albicans (fungal strain) [10]. In addition, Martina's group employed intelligent superabsorbent polymeric hydrogels, utilizing them as both a water reservoir and a supplier of mineral and organic nutrients with the aim of enhancing soil quality. In this case, the incorporation of NPK fertilizer reinforced the flexibility of the hydrogels, whereas the inclusion of lignohumate induced a shift in their rheological properties towards a more liquid-like behavior [11]. In light of the growing interest and attention toward hydrogel-based smart soils and relevant sustainable devices for tackling agricultural and environmental challenges, we recommend considering the following perspectives of smart hydrogels: (i) enhancing moisture retention and irrigation efficiency, (ii) implementing controlled nutrient release mechanisms, (iii) integrating soil health monitoring and environmental protection, (iv) ensuring compatibility with soil microorganisms, and (v) exploring additional avenues [12].

In hydrogel-based smart devices, Hideaki and colleagues unveiled that poly(stearyl acrylate (SA)-co-methoxy poly(ethylene glycol) acrylate (MPGA)) hydrogels featuring XSA > 0.5 trigger a transition from crystalline to amorphous state, constituting a challenging shift from hardness to softness at approximately 40 °C [13]. These hydrogels exhibited remarkable volume stability, irrespective of temperature fluctuations. This distinct attribute led to the utilization of poly(SA-co-MPGA) hydrogel in the development of shape memory "devices", rendering them pliable and flexible at temperatures exceeding 40 °C, and stiffening when cooled below 37.5 °C. Hydrogels with a high water content (50~99%) are susceptible to freezing, leading to reduced flexibility at low temperatures, significantly restricting their utility in cold environments. To address this, Tang and coworkers employed butanediol (BD) and N-hydroxyethyl acrylamide (HEAA) monomers, both featuring a multi-hydrogen bond structure, to fabricate a LiCl/p(HEAA-co-BD) conductive hydrogel with anti-freezing properties [14]. They achieved this by strategically manipulating intermolecular and intramolecular hydrogel bonds within the crosslinking network, effectively inhibiting the formation of primary ice crystals. The designed smart hydrogel-based sensors enabled an excellent anti-freezing property with a low freeze point of −85.6 °C, while they maintained stretchability up to 400% with a tensile stress of ~450 kPa for human motion detection at −40 °C. Although smart hydrogel devices offer numerous advantages, including flexibility, biocompatibility, and sensitivity, they do have certain limitations, such as limited long-term stability, low gauge factor, and slow response time, among others. Continual research and innovations in this field are poised to tackle these hurdles, broadening the scope of potential applications, and enhancing the overall efficacy of hydrogel devices across diverse domains such as healthcare, environmental monitoring, and beyond.

3. Conclusions

We aspire for this Special Issue to offer readers insightful glimpses into the synthesis, characterization, and applications of both smart and tough hydrogels. Given the dynamic and rapidly advancing nature of this field, it is unfeasible to encompass every facet, particularly the recent breakthroughs from research groups not directly involved in this Special Issue. Undoubtedly, this burgeoning field will continue to flourish, drawing contributions from diverse disciplines such as chemistry, physics, materials science, and engineering. We also hope that readers will derive both enjoyment and inspiration from the diverse array of topics presented here, potentially propelling this field closer to the commercial significance of soft hydrogels in various promising fields.

Conflicts of Interest: The authors declare no conflict of interest.

References

1. Ohsedo, Y.; Ueno, W. Creation of Polymer Hydrogelator/Poly (Vinyl Alcohol) Composite Molecular Hydrogel Materials. *Gels* **2023**, *9*, 679. [CrossRef] [PubMed]
2. Mrohs, T.B.; Weichold, O. Hydrolytic Stability of Crosslinked, Highly Alkaline Diallyldimethylammonium Hydroxide Hydrogels. *Gels* **2022**, *8*, 669. [CrossRef] [PubMed]
3. Simeonov, M.; Apostolov, A.A.; Georgieva, M.; Tzankov, D.; Vassileva, E. Poly (acrylic acid-co-acrylamide)/Polyacrylamide pIPNs/Magnetite Composite Hydrogels: Synthesis and Characterization. *Gels* **2023**, *9*, 365. [CrossRef] [PubMed]
4. Zhang, D.; Tang, Y.; Zhang, K.; Xue, Y.; Zheng, S.Y.; Wu, B.; Zheng, J. Multiscale bilayer hydrogels enabled by macrophase separation. *Matter* **2022**, *6*, 1484–1502. [CrossRef]
5. Zhang, D.; Tang, Y.; He, X.; Gross, W.; Yang, J.; Zheng, J. Bilayer Hydrogels by Reactive-Induced Macrophase Separation. *ACS Macro Lett.* **2023**, *12*, 598–604. [CrossRef] [PubMed]
6. Ni, C.; Chen, D.; Yin, Y.; Wen, X.; Chen, X.; Yang, C.; Chen, G.; Sun, Z.; Wen, J.; Jiao, Y.; et al. Shape memory polymer with programmable recovery onset. *Nature* **2023**, 1–6. [CrossRef] [PubMed]
7. Suhail, M.; Chiu, I.H.; Lai, Y.R.; Khan, A.; Al-Sowayan, N.S.; Ullah, H.; Wu, P.C. Xanthan-Gum/Pluronic-F-127-Based-Drug-Loaded Polymeric Hydrogels Synthesized by Free Radical Polymerization Technique for Management of Attention-Deficit/Hyperactivity Disorder. *Gels* **2023**, *9*, 640. [CrossRef] [PubMed]
8. Wang, Y.; Peng, Z.; Zhang, D.; Song, D. Tough, Injectable Calcium Phosphate Cement Based Composite Hydrogels to Promote Osteogenesis. *Gels* **2023**, *9*, 302. [CrossRef] [PubMed]
9. Shi, T.; Xie, Z.; Mo, X.; Feng, Y.; Peng, T.; Song, D. Highly efficient adsorption of heavy metals and cationic dyes by smart functionalized sodium alginate hydrogels. *Gels* **2022**, *8*, 343. [CrossRef] [PubMed]
10. Crețu, B.E.B.; Rusu, A.G.; Ghilan, A.; Rosca, I.; Nita, L.E.; Chiriac, A.P. Cryogel System Based on Poly (vinyl alcohol)/Poly (ethylene brassylate-co-squaric acid) Platform with Dual Bioactive Activity. *Gels* **2023**, *9*, 174. [CrossRef] [PubMed]
11. Kratochvílová, R.; Kráčalík, M.; Smilková, M.; Sedláček, P.; Pekař, M.; Bradt, E.; Smilek, J.; Závodská, P.; Klučáková, M. Functional Hydrogels for Agricultural Application. *Gels* **2023**, *9*, 590. [CrossRef] [PubMed]
12. Zhang, D.; Tang, Y.; Zhang, C.; Huhe, F.N.U.; Wu, B.; Gong, X.; Chuang, S.S.; Zheng, J. Formulating zwitterionic, responsive polymers for designing smart soils. *Small* **2022**, *18*, 2203899. [CrossRef] [PubMed]

13. Tokuyama, H.; Iriki, R.; Kubota, M. Thermosensitive Shape-Memory Poly (stearyl acrylate-co-methoxy poly (ethylene glycol) acrylate) Hydrogels. *Gels* **2023**, *9*, 54. [CrossRef] [PubMed]
14. Dai, B.; Cui, T.; Xu, Y.; Wu, S.; Li, Y.; Wang, W.; Liu, S.; Tang, J.; Tang, L. Smart antifreeze hydrogels with abundant hydrogen bonding for conductive flexible sensors. *Gels* **2023**, *8*, 374. [CrossRef] [PubMed]

Disclaimer/Publisher's Note: The statements, opinions and data contained in all publications are solely those of the individual author(s) and contributor(s) and not of MDPI and/or the editor(s). MDPI and/or the editor(s) disclaim responsibility for any injury to people or property resulting from any ideas, methods, instructions or products referred to in the content.

Article

Creation of Polymer Hydrogelator/Poly(Vinyl Alcohol) Composite Molecular Hydrogel Materials

Yutaka Ohsedo [1,*] and Wakana Ueno [2]

[1] Division of Engineering, Faculty of Engineering, Nara Women's University, Kitauoyahigashi-Machi, Nara 630-8506, Japan
[2] Faculty of Human Life and Environment, Nara Women's University, Kitauoyahigashi-Machi, Nara 630-8506, Japan
* Correspondence: ohsedo@cc.nara-wu.ac.jp

Citation: Ohsedo, Y.; Ueno, W. Creation of Polymer Hydrogelator/Poly(Vinyl Alcohol) Composite Molecular Hydrogel Materials. *Gels* 2023, 9, 679. https://doi.org/10.3390/gels9090679

Academic Editors: Dong Zhang, Jintao Yang, Xiaoxia Le and Dianwen Song

Received: 10 August 2023
Revised: 21 August 2023
Accepted: 22 August 2023
Published: 23 August 2023

Copyright: © 2023 by the authors. Licensee MDPI, Basel, Switzerland. This article is an open access article distributed under the terms and conditions of the Creative Commons Attribution (CC BY) license (https://creativecommons.org/licenses/by/4.0/).

Abstract: Polymer hydrogels, including molecular hydrogels, are expected to become materials for healthcare and medical applications, but there is a need to create new functional molecular gels that can meet the required performance. In this paper, for creating new molecular hydrogel materials, the gel formation behavior and its rheological properties for the molecular gels composed of a polymer hydrogelator, poly(3-sodium sulfo-*p*-phenylene-terephthalamide) polymer (**NaPPDT**), and water-soluble polymer with the polar group, poly(vinyl alcohol) (**PVA**) in various concentrations were examined. Molecular hydrogel composites formed from simple mixtures of **NaPPDT** aqueous solutions (0.1 wt.%~1.0 wt.%) and **PVA** aqueous solutions exhibited thixotropic behavior in the relatively low concentration region (0.1 wt.%~1.0 wt.%) and spinnable gel formation in the dense concentration region (4.0 wt.%~8.0 wt.%) with 1.0 wt.% **NaPPDT** aq., showing a characteristic concentration dependence of mechanical behavior. In contrast, each single-component aqueous solution showed no such gel formation in the concentration range in the present experiments. No gel formation behavior was also observed when mixed with common anionic polymers other than **NaPPDT**. This improvement in gel-forming ability due to mixing may be due to the increased density of the gel's network structure composed of hydrogelator and **PVA** and rigidity owing to **NaPPDT**.

Keywords: polymer hydrogelator; molecular hydrogel; thixotropic behavior; poly(vinyl alcohol); composite

1. Introduction

Gels encompass old and new forms of substances and materials that are typically found in food and other aspects of life [1]. These gel-like substances are known to be formed from various materials such as organic polymers and clay minerals; such substances that are formed from polymers are typically regarded as polymer gels [2,3]. Polymer gels contain solvents in the three-dimensional network structure of polymer chains, and there is interest in academic investigations aimed at elucidating the correlation between material properties and the polymer structure [2,3]. Conversely, by designing and controlling the interaction between polymer chains or solvent molecules inside a polymer gel, or by designing and controlling the network structure, it is possible to introduce various functionalities, including external stimulus responsiveness. Thus, there is interest in the development and application of intelligent functions such as sensors and actuators, and active research and development is underway [4]. In particular, polymer hydrogels, which are hydrogels and polymer gels, are expected to have an affinity with living organisms, as their structure is similar to that of cells, the extracellular matrix, and other biomaterials because the living organism itself can be considered a hydrogel. For these reasons, polymer hydrogels are being actively investigated as polymeric biomaterials that can be used as cell scaffold materials, drug delivery systems, or artificial organs, and their development is attracting attention from both the basic and applied perspectives [4–7]. In applications other than biomaterials, polymer gels have attracted attention as a form of functional polymeric material, and various research and development efforts are underway for their practical applications. For example, gel actuators that realize macroscopic movements using polymer hydrogels are attracting attention as a material that can lead to the creation of artificial muscles made from polymer hydrogels in the future because of the smooth and soft movements derived from gel-like substances and the expected biocompatibility of polymer hydrogels which is expected to be applied in the medical field [8]. As described above, polymer gels, especially polymer hydrogels, are one of the material forms and gel materials that have been actively focused on in recent years for the creation of new materials in terms of both basic science and applications. Among various polymer gels, molecular gels composed of polymer gelators have received remarkable attention for research and application due to their ease of acquisition, as gels can be reproducibly obtained simply by standing a solution obtained from thermal dissolution in a solvent or room-temperature dissolution [9–14].

We have focused on molecular gels that form self-assembled or cohesive fiber structures and are physically crosslinked by reversible interactions between molecules, and as part of our research on the fabrication of new molecular gel materials, we have conducted a series of studies on the function of water-soluble aromatic polyamide poly(3-sodium sulfo-*p*-phenylene-terephthalamide) (**NaPPDT**) [15] as a polymer hydrogelator. The hydrogel obtained from this gelator possesses a thixotropic property [16–18], which is of considerable interest as a required essential property for creams and ointments, as it exhibits a reversible change from a sol state to a gel state when subjected to an external mechanical force. Furthermore, this polymer hydrogelator is a thixotropic composite gel material that serves as a matrix for various organic and inorganic materials, such as inorganic nanosheet Laponite [19] and water-dispersible polyanilines [20,21], both of which have potential applications as a base material for ointments in the healthcare fields [22–25].

Herein, to fabricate novel molecular gel materials using polymer gelators as a matrix in healthcare fields, a new composite molecular hydrogel is fabricated by mixing a water-soluble polymer with **NaPPDT** (Figure 1). Poly(vinyl alcohol) (**PVA**) was selected as the water-soluble polymer to be mixed, as it is an electrostatically neutral non-ionic water-soluble polymer that is not expected to form water-insoluble polyionic complexes with **NaPPDT** containing a sodium sulfonate salt moiety, and it is a proven compound as a hydrophilic component in numerous polymer composites [26,27]. In the fabrication of composite gels by mixing polymers with gelling agents, the behavior of gel formation may be tuned by the molecular weight of the polymers. In this study, simple mixing of **NaPPDT** and **PVA** aqueous solutions at room temperature resulted in composite molecular gels that exhibit different mechanical properties compared to **NaPPDT** molecular hydrogels. This study revealed a thixotropic hydrogel material with a novel polymeric network structure, which is not observed in a single system of the same concentration region and is inferred to have a new polymeric network structure, as shown below.

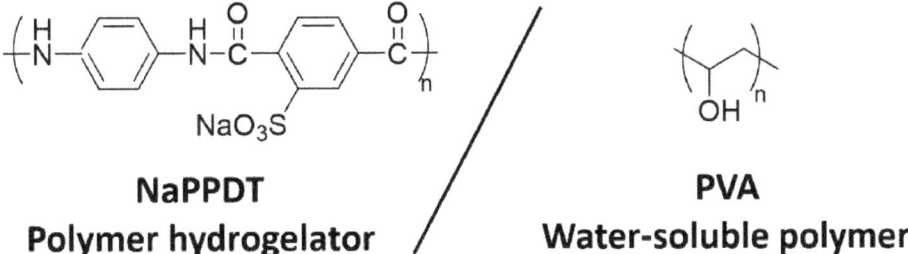

Figure 1. Chemical structures of polymer hydrogelator **NaPPDT** and water-soluble polymer **PVA**.

2. Results and Discussion

First, the mixing of the prepared **NaPPDT** aqueous solutions with **PVA** aqueous solutions at room temperature by vortex mixer and the resulting mixed solution were observed (Table 1 and Figure 2). The polymer hydrogelator **NaPPDT** was a sample obtained by polycondensation, according to a previous study [15]. The **NaPPDT** solution is the liquid state at 0.5 wt.%, but at 1.0 wt.%, the solution has good thixotropic properties with a recovery time to the gel state within 1 min after making sol by applying external force. The resulting mixed solution is a light-yellow molecular hydrogel with good thixotropic properties and a recovery time to gel within 1 min after solvation under external force. As **NaPPDT** at concentrations above 2.0 wt.% does not flow by pressing and requires time to mix with other components, aqueous solutions with **NaPPDT** concentrations of 0.5 wt.% (liquid) and 1.0 wt.% (gel) were used for mixing with the aqueous **PVA** solution. The adjusted aqueous **PVA** solution became viscous at 8.0 wt.%, and as mixing with aqueous **NaPPDT** solution by simple vortex mixing would be difficult at higher concentrations, a concentration series was prepared with the highest concentration of 8.0 wt.% for the **PVA** solutions. Aqueous **NaPPDT** solutions (which form a gel at ≥1.0 wt.%) and aqueous **PVA** solutions (which were in a solution state with no gel formation even at 8.0 wt.% for both types of **PVA** described below) were mixed at a weight ratio of 1:1, and the gel-forming ability of the mixtures was evaluated (low and high molecular weight PVAs were designated **PVA-L** and **PVA-H**, respectively, as shown in Table 1); the mixture of 1.0 wt.% **NaPPDT** formed a gel in both combinations, although the concentration of **NaPPDT** after mixing was 0.5 wt.%, which is not gel-forming, indicating that the complexes were formed by the interaction of **PVA** with **NaPPDT**, which is involved in gel formation. The resulting gel could be extracted and showed thixotropic properties.

Table 1. Gelation behavior of mixed solutions of **NaPPDT** aq. and **PVA** aq. (mixing at 1:1 by weight ratio).

	NaPPDT 0.1 wt.%	NaPPDT 0.5 wt.%	NaPPDT 1.0 wt.%
PVA-L 0.1 wt.%	L [1]	L	G [2]
PVA-L 0.5 wt.%	L	L	G
PVA-L 1.0 wt.%	L	GD [3]	G
PVA-L 2.0 wt.%	–	–	G
PVA-L 4.0 wt.%	–	–	G
PVA-L 8.0 wt.%	–	–	G
PVA-H 0.1 wt.%	L	L	G
PVA-H 0.5 wt.%	L	GD	G
PVA-H 1.0 wt.%	L	GD	G
PVA-H 2.0 wt.%	–	–	SG [4]
PVA-H 4.0 wt.%	–	–	SG
PVA-H 8.0 wt.%	–	–	SG

[1] L: liquid state. [2] G: gel state. [3] WG: gel that drips off after 5 min. [4] SG: spinnable gel.

To determine whether hydrogel formation by mixing aqueous **NaPPDT/PVA** systems in this low concentration range is a common phenomenon observed in the mixing of other anionic polymer/**PVA** systems other than **NaPPDT**, with other sodium sulfonates or phosphoric acids as side-chain polar groups of each of the polymers (1.0 wt.%) used, i.e., poly(sodium 4-styrenesulfonate) (two polymers with different molecular weights), chondroitin sulphate sodium salt, and deoxyribonucleic acid sodium salt, 1.0 wt.% aqueous polymer solutions were mixed with 1.0 wt.% aqueous **PVA** solutions and observed (Table 2). The results showed that the mixture was a clear solution and did not form a gel, indicating that the thixotropic gel formation due to earlier mixing in the low concentration range is a characteristic result of the **NaPPDT/PVA** complex. This is presumably because a certain degree of main chain rigidity is beneficial for gel formation at low concentrations, and other nonconjugated polymers with sodium sulfonate or phosphoric acid as side-chain polar groups lack rigidity or concentration. The mixing of higher concentrations of **PVA** solutions from 2.0 wt.% to 8.0 wt.% with a 1.0 wt.% **NaPPDT** solution was then investigated (Table 1). The results showed that, as with the mixing in the low concentration range, gels were obtained by mixing, but when the **PVA** concentration was above 4.0 wt.%, the gels obtained showed spinnability not seen in **PVA** alone systems of the same concentration, indicating that spinnable hydrogels were obtained (Figure 2h). These results also suggest that complexation of **PVA** and **NaPPDT** by mixing is involved in gel formation and that complexation by mixing is involved in the development of the spinnability of the gels obtained.

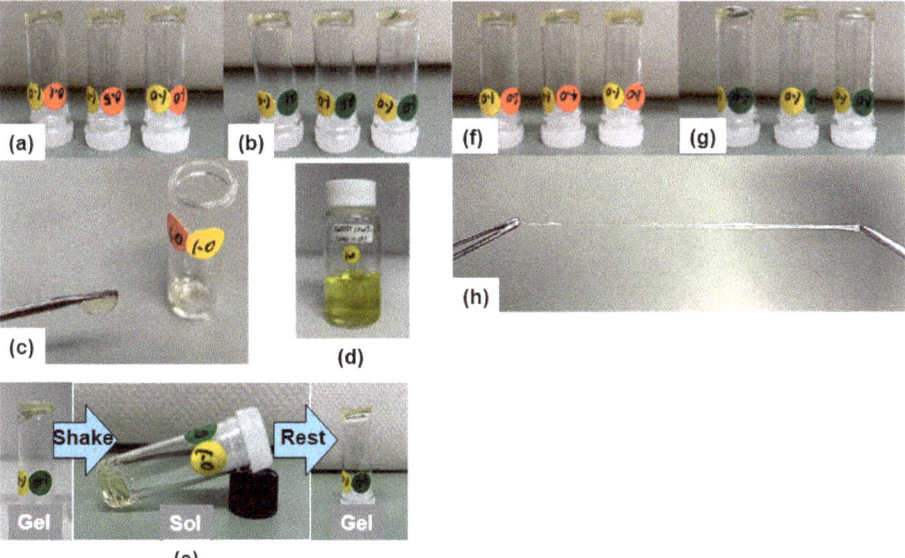

Figure 2. Gelation and thixotropic behavior of mixed composite **NaPPDT/PVA** 1/1 (w/w) molecular gel systems (concentration ratios shown as "wt.%/wt.%"): (**a**) **NaPPDT/PVA-L** system, left: 1.0/0.1, middle: 1.0/0.5, right: 1.0/1.0; (**b**) **NaPPDT/PVA-H** system, left: 1.0/0.1, middle: 1.0/0.5, right: 1.0/1.0; (**c**) thixotropic behavior of the scooped composite gel **NaPPDT/PVA-L** 1.0/1.0; (**d**) **NaPPDT** 1.0 wt.%; (**e**) **NaPPDT/PVA-H** 1.0/1.0 (shaken and sol, then left to stand for 1 min and recovered to gel state); (**f**) **NaPPDT/PVA-L**; left: 1.0/2.0; middle: 1.0/4.0; right: 1.0/8.0; (**g**) **NaPPDT/PVA-H**; left: 1.0/2.0; middle: 1.0/4.0; right: 1.0/4.0; right: 1.0/8.0; (**h**) pulled spinnable gel obtained from **NaPPDT/PVA-H** 1.0/8.0.

Table 2. Gelation behavior of aqueous anionic polymer solutions mixed with aqueous **PVA** solutions (mixing at 1:1 by weight ratio).

	PVA-L 1.0 wt.%	PVA-H 1.0 wt.%
NaPPDT 1.0 wt.%	G [1]	G
poly(sodium 4-styrenesulfonate) (M_w 70,000) 1.0 wt.%	L [2]	L
poly(sodium 4-styrenesulfonate) (M_w 1,000,000) 1.0 wt.%	L	L
Chondroitin sulfate sodium salt 1.0 wt.%	L	L
Deoxyribonucleic acid sodium salt 1.0 wt.%	L	L

[1] G: gel state. [2] L: liquid state.

To quantitatively evaluate the composite gel formation described above, dynamic viscoelasticity measurements (strain dispersion measurements) were performed using a rheometer, and the results are shown in Figure 3. In both systems, the composite gels were gels that transitioned from $G' > G''$ (gel) to $G' < G''$ (sol) [28] and had a lower modulus than **NaPPDT** alone at all **PVA** concentrations, and they are softer than **NaPPDT** alone at all **PVA** concentrations. The elastic modulus tended to increase with increasing **PVA** concentration, and the transition strain increased compared with the **NaPPDT** alone system, making the gels more resistant to deformation. In contrast, the **NaPPDT/PVA-H** system produced a stable gel with a higher modulus than the **PVA-L** system (less blurring of the measurement plots). The elastic modulus in the region above 100% strain was greater than that of the **NaPPDT** alone system. This shows that the higher molecular weight of **PVA** may have improved the gel network to the extent that it exhibited a high elastic modulus, resulting in the development of towing properties from gel formation. There was also an interesting trend toward higher modulus but lower transition strain at **PVA** concentrations that resulted in spinnable gels. The details of this phenomenon are a subject for further research, but it may be related to the gel's spinnability and ability to expand.

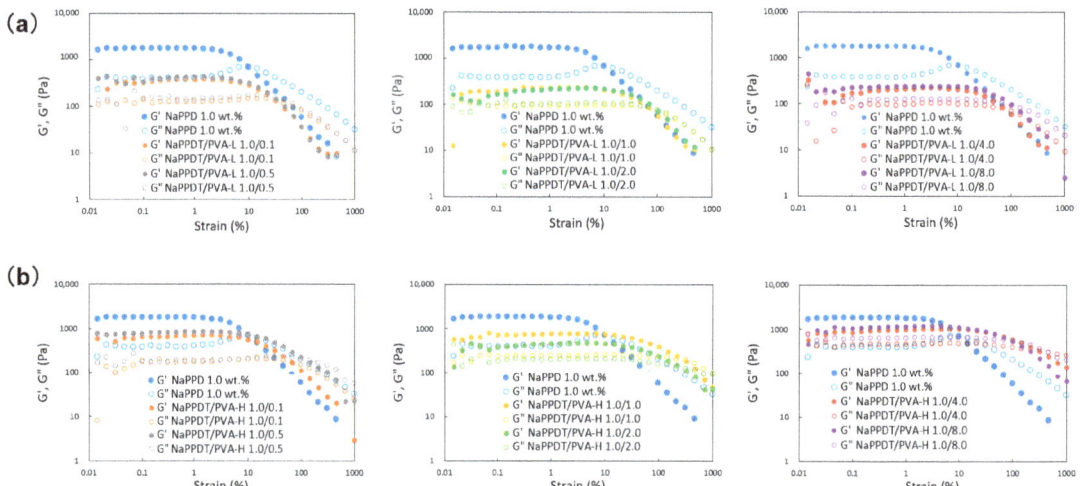

Figure 3. Dynamic rheological properties of the mixed composite **NaPPDT/PVA** 1/1 (w/w) molecular gel systems on strain sweep: (**a**) **NaPPDT/PVA-L** systems; (**b**) **NaPPDT/PVA-H** systems.

Next, the thixotropic properties of the obtained composite gels were quantitatively evaluated using dynamic viscoelasticity measurements. The results shown in Figure 4 indicate that both composite hydrogels in both **PVA-L** and **PVA-H** systems exhibit reversible recovery from gel to sol after significant deformation. The degree of recovery, as determined by comparing the modulus of elasticity, showed that they had almost returned to their pre-deformation state. A trend from $G' > G''$ to $G' \geq G''$ was observed after gel recovery as the **PVA** solution concentration increased in both systems. The scatter of the elastic modulus plots and the shape stability indicate that the high molecular weight **PVA-H** is more stable than **PVA-L**. This suggests that the high molecular weight **PVA-H** has an advantage in recovering to the gel state after large deformation. This may indicate that the presence of high molecular weight **PVA** is more effective in inter-network linkages in the recovery behavior and macroscopic recovery of the gel state. The reason for this could be that higher molecular weight, longer molecular chain lengths, and larger average radii of polymer chain filaments are more favorable for inter-network linkages compared to lower molecular weight polymers. However, although the **PVA-H** composite gels recovered repeatedly, at higher **PVA** concentrations the distance between G' and G'' was almost the same as at low **PVA** concentration. This is probably due to the predominance of the **NaPPDT** property at low **PVA** concentrations, but the predominance of the liquid property of **PVA** as the **PVA** concentration increases; spinnable gels were obtained at high **PVA** concentrations, but the spinnable property is close to liquid and gel may be important. This may be because the plots were stable in a stable gel state, but as the **PVA** concentration increased and the liquid properties appeared, the plots became unstable and blurred. This trend was more pronounced for the low molecular weight **PVA-L** composite gel compared to **PVA-H**, which may be related to the fact that low molecular weight **PVA-L** is less favorable for recovery to a gel after large deformation.

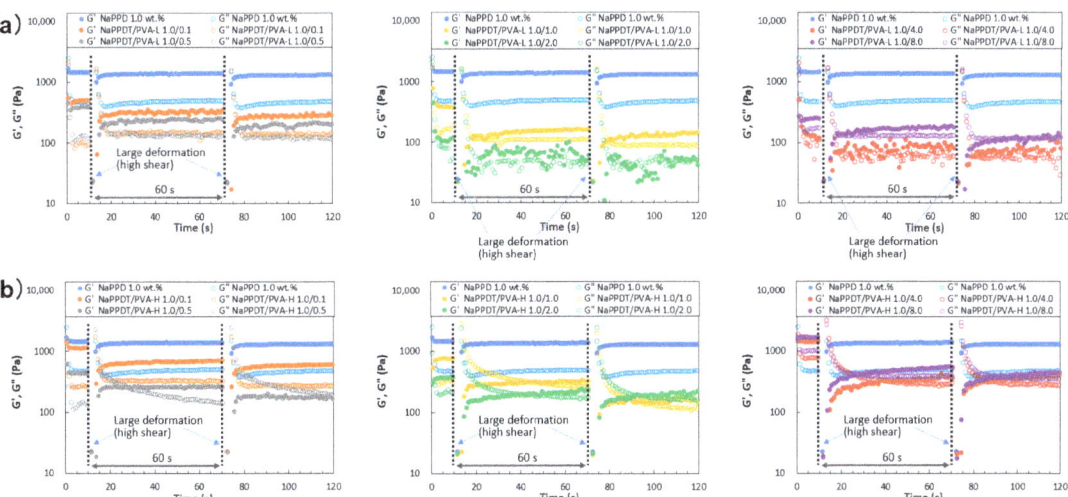

Figure 4. Thixotropic behavior of the mixed composite **NaPPDT/PVA** 1/1 (w/w) molecular gel systems (concentration ratios shown as wt.%/wt.%): (**a**) **NaPPDT/PVA-L** systems; (**b**) **NaPPDT/PVA-H** systems.

The falling ball method evaluated the transition temperature from gel to sol during the temperature increase to evaluate the thermal stability of the mixed composite molecular hydrogels. The results (Table 3) showed that the transition temperature of the hydrogels ranged from 40 °C to 60 °C and that they maintained their gel state at about 37 °C, the body temperature of the human body. This suggests that these hydrogels have potential for healthcare applications involving contact with the human body, such as ointment base materials. These results also showed that the transition temperature tended to be lower when the **PVA** had a lower molecular weight. This corresponds to the lower elastic modulus and softer properties of the lower molecular weight **PVA** in the rheometry measurements, which is thought to facilitate the transition due to increased temperature. In addition, for all **PVAs**, the transition temperature tended to decrease when the **PVA** concentration was 2.0 wt.% or higher. This corresponds to the result that, in rheometry measurements, the higher the **PVA** concentration, the more spinnable the gel becomes, and the closer it is to a liquid, the softer and easier it is to flow, which is thought to facilitate the transition when the temperature rises.

Scanning electron microscopy (SEM) observations of the freeze-dried xerogels were used to examine the internal microstructure of the composite gels in the μm to nm range and the surface topography of the constituent elements. From the SEM image in Figure 5, the xerogel of **NaPPDT** alone appeared to be a folded aggregate of several μm wide bands of material that were partially fused together. The dried **PVA** solution also appeared to have overlapping fibrous components of tens of nm width and appeared to be plate-like. The mixed composite xerogels obtained by mixing these materials were found to have finer constituents than the raw material components, particularly the **NaPPDT** xerogels. This suggests that the **NaPPDT** mixed with **PVA** interacted with each other to unwind the aggregates of the **NaPPDT** μm diameter material band into fibrous aggregates of smaller diameters and thicknesses, forming a new fine dense network of which it was a component.

This qualitative improvement in the composite gel network due to mixing may be the reason for the improved mechanical properties as measured by dynamic viscoelasticity described above. Although there is a concern that observation of such dried xerogel samples will show artefacts not present in the original gel-like samples due to shape changes during the drying process, it is thought that even if there are changes during the drying process, it is certain that the components have become smaller due to the mixing process as shown in our previous studies [19–21].

Table 3. Gel to sol transition temperature of the hydrogels results by falling-ball method.

Mixed Composite Sample	Temperature of Gel to Sol (°C)
PVA-L 0.1 wt.%/NaPPDT 1.0 wt.%	50 [1]
PVA-L 0.5 wt.% /NaPPDT 1.0 wt.%	50
PVA-L 1.0 wt.% /NaPPDT 1.0 wt.%	50
PVA-L 2.0 wt.% /NaPPDT 1.0 wt.%	40
PVA-L 4.0 wt.% /NaPPDT 1.0 wt.%	40
PVA-L 8.0 wt.% /NaPPDT 1.0 wt.%	40
PVA-H 0.1 wt.% /NaPPDT 1.0 wt.%	60
PVA-H 0.5 wt.% /NaPPDT 1.0 wt.%	60
PVA-H 1.0 wt.% /NaPPDT 1.0 wt.%	61
PVA-H 2.0 wt.% /NaPPDT 1.0 wt.%	60
PVA-H 4.0 wt.% /NaPPDT 1.0 wt.%	57
PVA-H 8.0 wt.% /NaPPDT 1.0 wt.%	57

[1] The evaluation was performed by increasing the substrate plate temperature from 25 °C at 1 °C/min.

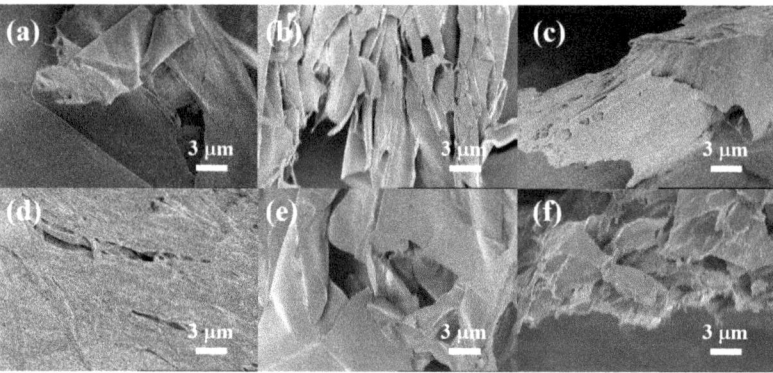

Figure 5. SEM images of the mixed composite and single xerogels and dried sample of **PVA** obtained from freeze-drying of hydrogels and **PVA** aqueous solution: (**a**) xerogel obtained from 1.0 wt.% NaPPDT aq.%; (**b**) freeze-dried sample obtained from 8.0 wt.% **PVA-H** aq.; (**c**) xerogel obtained from **NaPPDT/PVA-L** 1.0 wt./1.0 wt.%; (**d**) xerogel obtained from **NaPPDT/PVA-H** 1.0 wt./1.0 wt.%; (**e**) xerogels obtained from **NaPPDT/PVA-L** 1.0 wt./8.0 wt.%; (**f**) xerogels obtained from **NaPPDT/PVA-H** 1.0 wt./8.0 wt.%.

To see the intermolecular interactions between the components in the composite gels, the attenuated total reflectance Fourier transform infrared spectroscopy (attenuated total reflectance (ATR)–FTIR) absorption spectra of the xerogel and the dry samples were evaluated. As depicted in Figure 6a,c, other than the addition of the individual components, no new absorption bands or significant shifts in the absorption bands were observed in the xerogel of the composite gel in the wavenumber region of the stretching vibration of the hydroxyl and sodium sulfonate moieties (3000–2600 cm^{-1}). No significant changes were also observed in the absorption region of the amide bonding sites of **NaPPDT**. This is presumably due to the lower concentration of sites contributing to intermolecular interactions involved in mesh formation compared with the concentration of sites involved in absorption bands due to inter- and intramolecular interactions involved for a single component. A schematic illustration of the mesh structure before and after mixing, combining these results with those obtained from rheometry and SEM images, is shown in Figure 6c. As shown in Figure 6c, the mesh of the **NaPPDT** gel was loosened by the addition of the second component, **PVA**, but the mesh was maintained. Further studies on the mesh structure in gels using various composite samples will be necessary in the future.

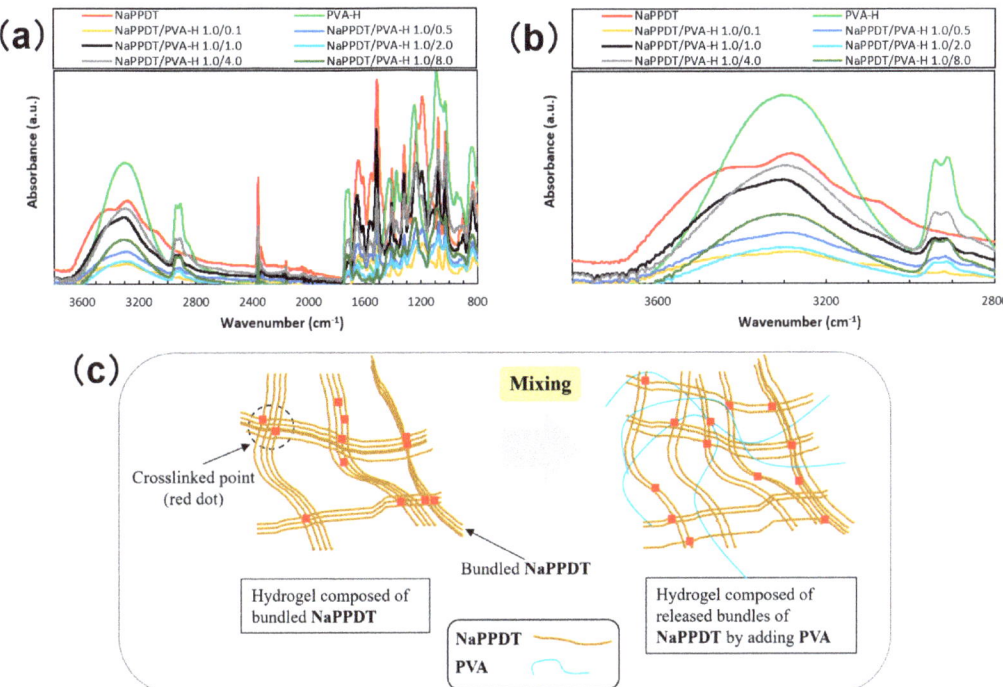

Figure 6. ATR-FTIR spectra of xerogels obtained from **NaPPDT** and composite molecular hydrogels and freeze-dried sample of **PVA-H** (concentration ratios shown as wt.%/wt.%): (**a**) spectra of the entire measurement region; (**b**) spectra of hydroxyl groups and −ONa unit in the stretching vibration region; (**c**) schematic illustrations of a mixed composite hydrogel.

Finally, in order to see the potential application of spinnable hydrogel, an attempt was made to obtain fibrous samples from samples in which spinnable hydrogel was formed by mixing **NaPPDT/PVA-L** 1.0 wt.%/8.0 wt.%. For the sample after mixing in the vial, the edges of the gel were pinched with tweezers and pulled to obtain fibrous material. The fibrous material was dried at 60 °C to obtain dried fibers. A photograph, images of SEM, and polarized light microscope of the dried composite gel fibers are shown below. As shown in this Figure 7, straight fibers were obtained in μm order. Polarized light microscopy (under crossed Nicols conditions) showed that the area where the fibers were present appeared bright, indicating that the fibers were oriented or anisotropic. This is thought to be due to the fibers being oriented as a result of drawing to form the fibers and the orientation or anisotropy because **NaPPDT** exhibits lyotropic liquid crystallinity in the mixed concentration range. Thus, it was found that oriented fibers could be obtained from the composite gel. However, as we are still investigating experimental conditions to obtain fiber samples with constant diameter, mechanical property tests including tensile tests of the fibers will be considered in the future.

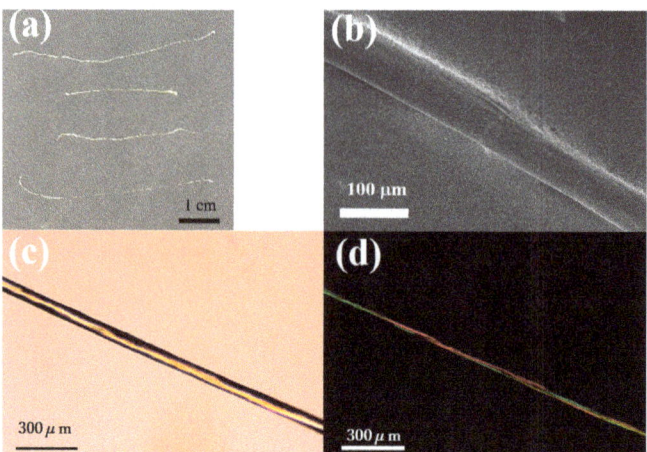

Figure 7. Photograph and images of SEM and polarized light microscope image of the composite gel fiber of **NaPPDT/PVA-H** 1.0 wt.%/8.0 wt.%: (**a**) a photograph of the fiber; (**b**) a SEM image of the fiber; (**c**) a polarized light microscope image of the fiber (without a polarizer); (**d**) a polarized light microscope image of the fiber (under crossed Nicols conditions).

3. Conclusions

A gel-like substance with thixotropic properties was obtained by mixing the polymer hydrogelator **NaPPDT** with the water-soluble polymer **PVA** at low concentrations, where the single component of composites would be a liquid. This is possible with the combination of **NaPPDT** and **PVA** and is a special phenomenon that is not observed with mixtures of aqueous anionic polymer solutions with other sodium sulfonate or phosphonic acid side chains and aqueous **PVA** solutions. In addition, increasing the **PVA** concentration in the composite gel to a high concentration of 2.0 wt.% or more, gel stabilization and the development of spinnable gel properties were observed due to an increase in the gel-sol transition strain, and an increase in the elastic modulus was observed by increasing the molecular weight of **PVA** in the composite gel. These observations can be attributed to the qualitative improvement in the network structure of the gel due to the increased concentration and high molecular weight. Thus, despite the dilution of gelator and **PVA** concentrations when mixing aqueous polymer hydrogelator and **PVA** solutions, composite molecular gel materials with improved mechanical properties could be produced. As this gel material has been shown to be a matrix for functional materials such as **NaPPDT** and **PVA**, respectively, it can be expected to be a potential candidate for a base material for ointments in the healthcare field as a new gel-like matrix.

4. Materials and Methods

The polymer hydrogelator **NaPPDT** (M_n = 10,000) was prepared according to the previous literature [15] by polycondensation of phenylenediamine and sodium 2-sulfotrerephthalate with LiCl in *N*-methyl-2-pyrrolidone using the phosphorylation method. Pure water was deionized with an Elix UV 3 Milli-Q integral water purification system (Nihon Millipore K.K., Tokyo, Japan). Low molecular weight poly(vinyl alcohol) (average mol wt. 30,000–70,000, 87–90% hydrolyzed), low molecular weight poly(vinyl alcohol) (average M_w 146,000–186,000, 87–89% hydrolyzed) poly(sodium 4-styrenesulfonate) (average M_w 70,000), and poly(sodium 4-styrenesulfonate) (average M_w 1,000,000) were purchased from Sigma-Aldrich Japan (Merck KGaA, Darmstadt, Germany) and used as received. Chondroitin sulfate sodium salt and deoxyribonucleic acid sodium salt were purchased from Tokyo Chemical Industry Co., Ltd. and used as received. All other chemicals were obtained from Wako Pure Chemical Industries, Ltd., Tokyo, Japan, and they were used without purification.

The preparation of **NaPPDT/PVA** composite molecular hydrogels was done as follows: at first, 0.5 wt.% and 1.0 wt.% **NaPPDT** aqueous solutions were made by mixing **NaPPDT** solid and pure water and rested for one day at room temperature (the 0.5 wt.% aqueous solution was liquid, and the 1.0 wt.% aqueous solution was gel). Then **PVA** aqueous solutions of various concentrations obtained by dissolving at 90 °C for one day were added to the **NaPPDT** aqueous solutions at 1:1 by weight ratio at room temperature and mixed by use of a vortex genie (Scientific Industries, Inc., Bohemia, New York, NY, USA). Before measurements, the mixed composite hydrogels were rested for 30 min at room temperature.

Gelation and thixotropic properties were determined using the vial inversion method. The vial inversion method judges a mixture as a gel if it does not fall out of the vial containing the mixture when the vial is inverted, and as a sol if it does. The vial inversion method was performed five minutes after the mixture was mixed to detect gelation. The gel was also judged to have returned to gel if the contents did not fall out when the vial was inverted again, after the vial had been left standing following the application of external mechanical force by a vortex mixer to the vial containing the gelled substance, and the gel was judged to have thixotropic properties. The falling-ball method was used to evaluate the change in state from gel to sol with increasing temperature. A 1 mm diameter, 7 mg weight SUS ball was gently placed on the top of the hydrogel in a vial, and the vial was placed on a cool plate CP-085 (Scinics Corporation, Tokyo, Japan) and wrapped with insulation (absorbent cotton), and the temperature of the base plate of the cool plate was raised from 25 °C at 1 °C/min for evaluation. In this study, we employed a Leica ML9300 polarized optical microscope (MEIJI TECHNO CO., LTD., Saitama, Japan) with crossed Nicols to conduct polarized light microscopy observations on composite hydrogels. For SEM image measurements, a JSM-6700FN scanning electron microscope (JEOL Ltd., Tokyo, Japan) operating at 1.0 keV was used. The freeze-dried xerogel samples were carefully positioned on a conductive tape situated on the brass SEM stage. Prior to imaging, a 10 nm thick of Pt was applied to the samples using a sputtering technique to enhance their electrical conductivity. Dynamic rheological measurements for samples were performed using an MCR 302e rheometer (Anton Paar Japan K.K., Tokyo, Japan) with a parallel plate set at a gap of 0.50 mm (8 mm diameter) at 25 °C. Frequency sweeps were conducted with a strain amplitude (γ) of 0.01%, and strain sweeps were carried out at a constant angular frequency of 1 rad s^{-1}. For the repeated step-shear measurements, a small strain with an amplitude of 0.01% and a frequency of 1 Hz was applied, followed by a large strain with a shear rate of 3000 s^{-1} for 0.1 s. Furthermore, ATR–FTIR spectra were recorded using a FTIR6600 spectrometer (JASCO Corporation, Tokyo, Japan) in conjunction with a single bounce diamond attenuated total reflectance (ATR) accessory.

Author Contributions: Conceptualization, Y.O.; investigation, Y.O. and W.U.; writing—original draft preparation, Y.O.; writing—review and editing, Y.O. and W.U.; supervision, Y.O.; funding acquisition, Y.O. All authors have read and agreed to the published version of the manuscript.

Funding: This research was partly funded by JSPS KAKENHI, grant numbers 15K05610 and 19K05634.

Institutional Review Board Statement: Not applicable.

Informed Consent Statement: Not applicable.

Data Availability Statement: Not applicable.

Conflicts of Interest: The authors declare no conflict of interest.

References

1. Horkay, F.; Douglas, J.F.; Del Gado, E. (Eds.) *Gels and Other Soft Amorphous Solids*; ACS Symposium Series 1296; American Chemical Society: Washington, DC, USA, 2018. [CrossRef]
2. DeRossi, D.; Kajiwara, K.; Osada, Y.; Yamauchi, A. (Eds.) *Polymer Gels*; Springer US: Boston, MA, USA, 1991.

3. Osada, Y.; Khokhlov, A.R. (Eds.) *Polymer Gels and Networks*; Marcel Dekker: New York, NY, USA, 2002.
4. Chelu, M.; Musuc, A.M. Polymer Gels: Classification and Recent Developments in Biomedical Applications. *Gels* **2023**, *9*, 161. [CrossRef] [PubMed]
5. Lee, K.Y.; Mooney, D.J. Hydrogels for Tissue Engineering. *Chem. Rev.* **2001**, *101*, 1869–1880. [CrossRef] [PubMed]
6. Guo, Y.; Bae, J.; Fang, Z.; Li, P.; Zhao, F.; Yu, G. Hydrogels and Hydrogel-Derived Materials for Energy and Water Sustainability. *Chem. Rev.* **2020**, *120*, 7642–7707. [CrossRef] [PubMed]
7. Correa, S.; Grosskopf, A.K.; Lopez Hernandez, H.; Chan, D.; Yu, A.C.; Stapleton, L.M.; Appel, E.A. Translational Applications of Hydrogels. *Chem. Rev.* **2021**, *121*, 11385–11457. [CrossRef] [PubMed]
8. Bonard, S.; Nandi, M.; García, J.I.H.; Maiti, B.; Abramov, A.; Díaz Díaz, D. Self-Healing Polymeric Soft Actuators. *Chem. Rev.* **2023**, *123*, 736–810. [CrossRef] [PubMed]
9. Guenet, J.-M. *Organogels Thermodynamics, Structure, Solvent Role, and Properties*; Springer International Publishing AG: Cham, Switzerland, 2016.
10. Weiss, R.G. (Ed.) *Molecular Gels, Structure and Dynamics*; The Royal Society of Chemistry: London, UK, 2018.
11. Wojtecki, R.J.; Meador, M.A.; Rowan, S.J. Using the Dynamic Bond to Access Macroscopically Responsive Structurally Dynamic Polymers. *Nat. Mater.* **2011**, *10*, 14–27. [CrossRef] [PubMed]
12. Miao, R.; Peng, J.; Fang, Y. Molecular Gels as Intermediates in the Synthesis of Porous Materials and Fluorescent Films: Concepts and Applications. *Langmuir* **2017**, *33*, 10419–10428. [CrossRef] [PubMed]
13. Weiss, R.G. Controlling Variables in Molecular Gel Science: How Can We Improve the State of the Art? *Gels* **2018**, *4*, 25. [CrossRef] [PubMed]
14. Cornwell, D.J.; Smith, D.K. Expanding the scope of gels—Combining polymers with low-molecular-weight gelators to yield modified self-assembling smart materials with high-tech applications. *Mater. Horiz.* **2015**, *2*, 279–293. [CrossRef]
15. Ohsedo, Y.; Oono, M.; Saruhashi, K.; Watanabe, H. A New Water-Soluble Aromatic Polyamide Hydrogelator with Thixotropic Properties. *RSC Adv.* **2015**, *5*, 82772–82776. [CrossRef]
16. Goodwin, J.; Hughes, R. *Rheology for Chemists: An Introduction*, 2nd ed.; Royal Society of Chemistry: Cambridge, UK, 2008.
17. Mewis, J.; Wagner, N.J. Thixotropy. *Adv. Colloid. Interface Sci.* **2009**, *147–148*, 214–227. [CrossRef] [PubMed]
18. Larson, R.G.; Wei, Y. A Review of Thixotropy and Its Rheological Modeling. *J. Rheol.* **2019**, *63*, 477–501. [CrossRef]
19. Ohsedo, Y.; Oono, M.; Saruhashi, K.; Watanabe, H.; Miyamoto, N. New Composite Thixotropic Hydrogel Composed of a Polymer Hydrogelator and a Nanosheet. *R. Soc. Open Sci.* **2017**, *4*, 171117. [CrossRef] [PubMed]
20. Ohsedo, Y.; Saruhashi, K.; Watanabe, H.; Miyamoto, N. Synthesis of an Electronically Conductive Hydrogel from a Hydrogelator and a Conducting Polymer. *New J. Chem.* **2017**, *41*, 9602–9606. [CrossRef]
21. Ohsedo, Y.; Sasaki, M. Polymeric Hydrogelator-Based Molecular Gels Containing Polyaniline/Phosphoric Acid Systems. *Gels* **2022**, *8*, 469. [CrossRef] [PubMed]
22. Dayan, N. Delivery System Design in Topically Applied Formulations: An Overview. In *Delivery System Handbook for Personal Care and Cosmetic Products, Technology, Applications, and Formulations*; Rosen, M.R., Ed.; William Andrew, Inc.: New York, NY, USA, 2005; pp. 101–118.
23. Sugibayashi, K.; Morimoto, Y. Transdermal Patches. In *Gels Handbook, The Fundamentals*; Osada, Y., Kajiwara, K., Fushimi, T., Irasa, O., Hirokawa, Y., Matsunaga, T., Shimomura, T., Wang, L., Ishida, H., Eds.; Elsevier Inc.: Amsterdam, The Netherlands, 2001; Volume 3, pp. 201–210.
24. Ohsedo, Y. Low-Molecular-Weight Gelators as Base Materials for Ointments. *Gels* **2016**, *2*, 13. [CrossRef] [PubMed]
25. Slavkova, M.; Tzankov, B.; Popova, T.; Voycheva, C. Gel Formulations for Topical Treatment of Skin Cancer: A Review. *Gels* **2023**, *9*, 352. [CrossRef] [PubMed]
26. Visakh, P.M.; Nazarenko, O.B. (Eds.) *Polyvinyl Alcohol-Based Biocomposites and Bionanocomposites*; John Wiley & Sons, Inc.: Hoboken, NJ, USA, 2023.
27. Palanichamy, K.; Anandan, M.; Sridhar, J.; Natarajan, V.; Dhandapani, A. PVA and PMMA nano-composites: A review on strategies, applications and future prospects. *Mater. Res. Express* **2023**, *10*, 022002. [CrossRef]
28. Kavanagh, G.M.; Ross-Murphy, S.B. Rheological Characterisation of Polymer Gels. *Prog. Polym. Sci.* **1998**, *23*, 533–562. [CrossRef]

Disclaimer/Publisher's Note: The statements, opinions and data contained in all publications are solely those of the individual author(s) and contributor(s) and not of MDPI and/or the editor(s). MDPI and/or the editor(s) disclaim responsibility for any injury to people or property resulting from any ideas, methods, instructions or products referred to in the content.

Article

Xanthan-Gum/Pluronic-F-127-Based-Drug-Loaded Polymeric Hydrogels Synthesized by Free Radical Polymerization Technique for Management of Attention-Deficit/Hyperactivity Disorder

Muhammad Suhail [1,*,†], I-Hui Chiu [1,†], Yi-Ru Lai [1], Arshad Khan [2], Noorah Saleh Al-Sowayan [3], Hamid Ullah [1] and Pao-Chu Wu [1,4,*]

1. School of Pharmacy, Kaohsiung Medical University, 100 Shih-Chuan 1st Road, Kaohsiung 80708, Taiwan; u112830001@kmu.edu.tw (I.-H.C.); u110830701@kmu.edu.tw (H.U.)
2. Department of Pharmaceutics, Faculty of Pharmacy, Khawaja Fareed Campus (Railway Road), The Islamia University of Bahawalpur, Punjab 63100, Pakistan; arshadpharma77@gmail.com
3. Department of Biology, College of Science, Qassim University, Buraydah 52377, Saudi Arabia; nsaoiean@qu.edu.sa
4. Drug Development and Value Creation Research Center, Kaohsiung Medical University, Kaohsiung 80708, Taiwan
* Correspondence: u108830004@kmu.edu.tw (M.S.); pachwu@kmu.edu.tw (P.-C.W.); Tel.: +886-7-3121101 (P.-C.W.)
† These authors contributed equally to this work.

Abstract: Smart and intelligent xanthan gum/pluronic F-127 hydrogels were fabricated for the controlled delivery of atomoxetine HCl. Different parameters such as DSC, TGA, FTIR, XRD, SEM, drug loading, porosity, swelling index, drug release, and kinetics modeling were appraised for the prepared matrices of hydrogels. FTIR confirmed the successful synthesis of the hydrogel, while TGA and DSC analysis indicated that the thermal stability of the reagents was improved after the polymerization technique. SEM revealed the hard surface of the hydrogel, while XRD indicated a reduction in crystallinity of the reagents. High gel fraction was achieved with high incorporated contents of the polymers and the monomer. An increase in porosity, drug loading, swelling, and drug release was observed with the increase in the concentrations of xanthan gum and acrylic acid, whereas Pluronic F-127 showed the opposite effect. A negligible swelling index was shown at pH 1.2 and 4.6 while greater swelling was observed at pH 7.4, indicating a pH-responsive nature of the designed hydrogels. Furthermore, a higher drug release was found at pH 7.4 compared to pH 1.2 and 4.6, respectively. The first kinetics order was followed by the prepared hydrogel formulations. Thus, it is signified from the discussion that smart xanthan gum/pluronic F-127 hydrogels have the potential to control the release of the atomoxetine HCl in the colon for an extended period of time.

Keywords: attention-deficit/hyperactivity disorder; atomoxetine; porosity; swelling; drug release

1. Introduction

Attention-deficit/hyperactivity disorder (ADHD) is one of the most common neurodevelopmental disorders in childhood and adolescence. This disorder affects between 2.2% and 17.8% of all school-aged children and adolescents. Different developmental deficits such as learning limitation, control of executive functions, as well as global impairments of social skills are associated with ADHD in children [1]. As early as 1902, "deficiencies in "volitional inhibition" and an excessive inability to pay attention for long periods of time" were seen with a group of restless children. In 1937, it was discovered that the levels of hyperactivity and behavioral issues could be reduced by amphetamine. In the 1950s, the term "minor brain damage" (MBD) was often used to describe these symptoms in children; however, in most cases, no evidence of neurological damage was identified. Until 1960, the disorder was known and labeled as ADHD and attention deficit disorder

(ADD). ADHD symptoms, inattention and impulsivity (and ADHD hyperactivity), have become attributed to a syndrome. The question of whether ADHD is a separate disorder or a continuum that serves as a risk factor for future adversities is one that is still being debated. With each new diagnostic system, the symptoms have changed, and the World Health Organization diagnostic system (ICD-10) now labels it as hyperkinetic disorder and the American Medical Association System (DSM-IV) describes it as ADHD, which look similar to each other [2]. Hence, different drugs such as atomoxetine HCl (ATMH) and a number of other drugs, including certain antidepressants and -agonists used off-label, have been added to the pharmacotherapy of ADHD [3]. The different recommended psychostimulants and other medications in the treatment of children and adolescents with ADHD are Methylphenidate, Ritalin Medikinet, Concerta, and the Amphetamine liquid, respectively. Cognitive behavioral therapy, neuropsychological treatment, noninvasive brain stimulation, and other multimodal treatments are the different non-pharmacological approaches that are used for the management of ADHD [4].

ATMH is approved in the US for the management of ADHD patients [5]. Its absorption occurs very rapidly after ingestion. The bioavailability is only affected slightly by the intake of food, so the drug can be taken independently of meals. Absolute bioavailability is commonly reached 63% after oral delivery, while the maximum plasma concentration is reached 1 to 2 h after ingestion. At therapeutic concentrations, the atomoxetine–albumin binding ratio is 98% in the plasma. High pre-systemic metabolism and albumin binding occur. The half-life is 4–5 h. Cytochrome P450-2D6 isoenzyme is responsible for the degradation of ATMH in the liver [5]. The available dosage form is a capsule with different doses of 10, 18, 25, 40, 60, 80, and 100 mg for adults and pediatrics, with a dose frequency of 40 mg PO once daily at first; 80 mg PO once daily or divided every 12 h after that; and maybe 100 mg if the desired reaction is not attained. ATMH presents some problems when ingested orally. It has a very short half-life and is absorbed very rapidly. Upset stomach, nausea, vomiting, constipation, loss of appetite, dry mouth, headache, lethargy, feeling drowsy, or weakness during the day are common adverse effects, while serious side effects like liver damage, elevated suicidal thoughts, angioedema, and heart problems are associated with a single-dose administration per day. Several researchers have prepared different carrier systems for ATMH delivery. Teaima and coworkers prepared solid lipid nanoparticles for the brain targeting of ATMH and sustained the release of ATMH for 8 h [6]. Mohanty et al. (2023) prepared ATMH-loaded nanostructured lipid carriers and demonstrated the sustained release of ATMH for 12 h [7]. Similarly, Stanojevic and coworkers prepared tablets for the sustained release of ATMH of up to 8 h [8]. Yet, further research work is needed to not only sustain the release of ATMH, but also combat its adverse effects following oral delivery. Hydrogel with unique properties has been used as one of the most suitable agents for controlled drug delivery systems [9]. Khalid et al. (2018) prepared polymeric hydrogels of chondroitin sulfate for the controlled release of loxoprofen [10]. Similarly, Malik and coworkers prepared chitosan/Beta-cyclodextrin-based hydrogels for acyclovir-controlled release [11].

In the current study, the authors report ATMH-loaded polymeric hydrogels of xanthan gum and pluronic F-127 fabricated by the free radical polymerization technique. The combination of natural polymer xanthan gum and synthetic polymer pluronic F-127 with synthetic monomer acrylic acid increased the pH sensitivity of the fabricated hydrogels. The mechanical strength and stability of the developed hydrogels were increased due to the intermixing of natural and synthetic contents, which not only enhanced the swelling and loading of drugs, but also sustained the release of ATMH for an extended period of time (96 h). Characterization techniques such as FTIR, TGA, DSC, XRD, and SEM were performed for the formulated hydrogels. Similarly, a set of studies including sol–gel analysis, porosity, swelling, drug loading and release, along with kinetics modeling have been performed for the fabricated networks of hydrogel. Gelation mechanism of prepared hydrogels is shown below in Scheme 1.

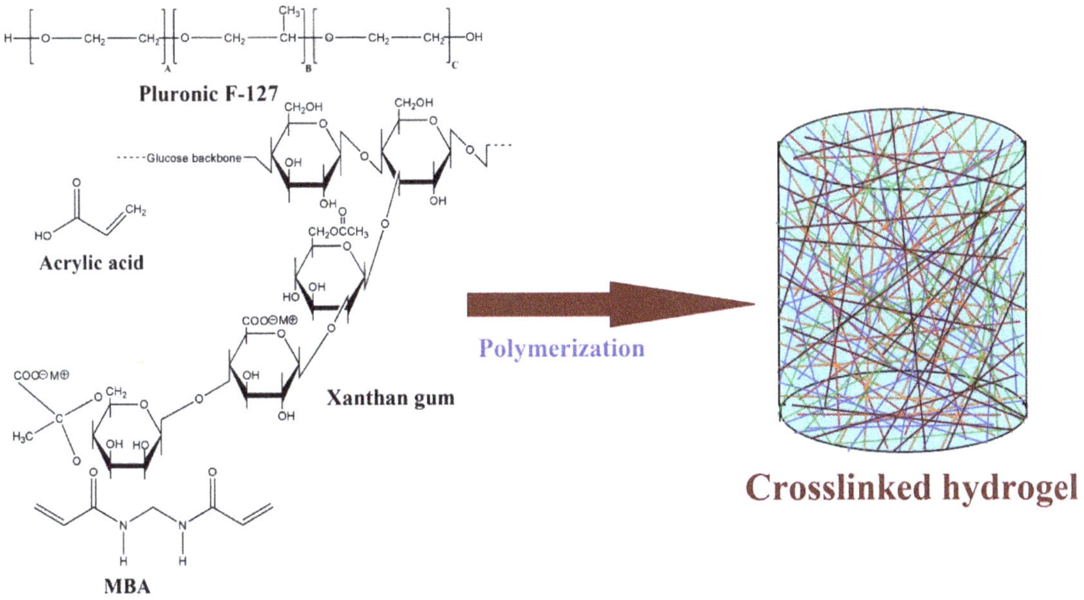

Scheme 1. Gelation mechanism of XG/PF-127 hydrogel.

2. Results and Discussion

2.1. FTIR Analysis

The nature of new bond formations and changes in the chemical structure of developed hydrogels were elucidated by FTIR analysis. The FTIR spectra of all components are indicated in Figure 1. XG indicated FTIR spectra by bands at 3280 and 1608 cm^{-1}, demonstrating hydrogen-bonded OH groups and COO− groups, whereas the bending of O–H and C–H was seen by peaks at 1018 and 1420 cm^{-1} [12,13], respectively. The FTIR spectra of PF-127 represent a C–O–C stretching vibration within the 1202–1002 cm^{-1} range. Likewise, the stretching vibration of COC and COC–CH$_2$ was seen by peaks at 1098 and 1059 cm^{-1}, whereas the CC–COC bond was confirmed by a peak at 1148 cm^{-1}. Two characteristic bands were observed at 1342 and 1468 cm^{-1}, indicating a CH$_2$ group. An absorption peak at 2918 cm^{-1} was assigned to the methyl group [14,15]. Similarly, the FTIR spectrum of Aa indicated a stretching vibration of –CH$_2$ by a broad band at 3014 cm^{-1}. Furthermore, the stretching vibration of C=O of the carboxylic acid was presented by a peak at 1702 cm^{-1} [16]. The FTIR spectra of the XG/PF127 hydrogel indicated prominent peaks of XG, PF127, and Aa. Certain new peaks were formed while some were modified, indicating the successful crosslinking among hydrogel components. The prominent bands of XG and PF127 were changed from 3280, 1608 cm^{-1}, and 2918, 1468 cm^{-1} to 2910, 1490, 2950, and 1505 cm^{-1}, respectively. Similarly, the position of certain peaks of Aa were also modified, like bands at 3014 and 1702 cm^{-1} which were moved to 3042 and 1685 cm^{-1}. This all demonstrates the synthesis of hydrogels. The FTIR spectrum of ATMH indicated the stretching vibration of the amino group (N–H) by a peak at 3270 cm^{-1}, while the stretching vibrations of CH$_2$ and COOH groups were observed at 2940 and 3322 cm^{-1}, respectively. The C=C ring stretching and the C–H bending of CH$_2$ groups were detected at 1598 and 1438 cm^{-1} [17]. After loading, a slight change was observed in the peaks of ATMH as 1438 and 2918 cm^{-1} were slightly shifted to 1412 and 2950 cm^{-1}, representing the successful loading of ATMH by the XG/PF-127 hydrogel [18].

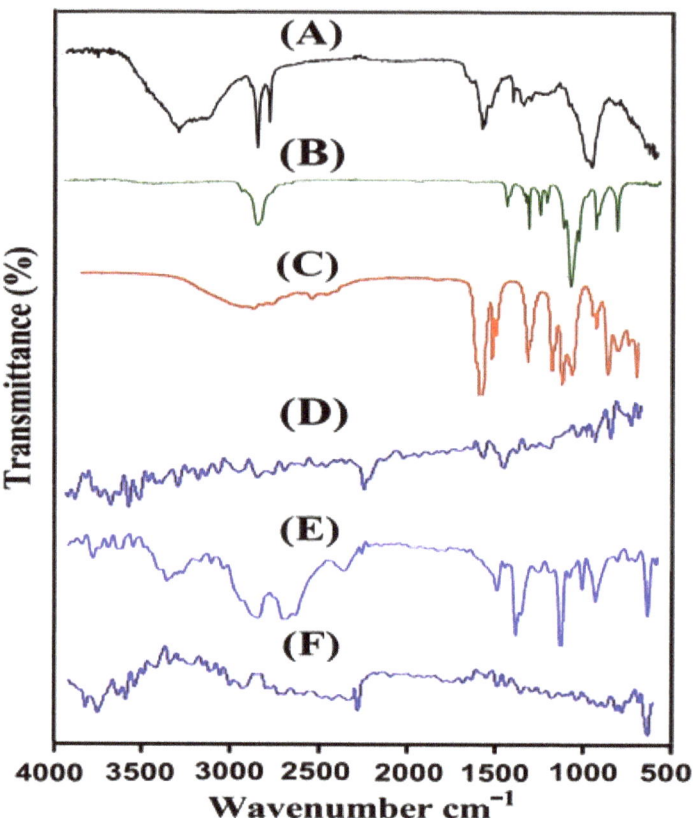

Figure 1. FTIR spectra of (**A**) XG, (**B**) PF-127, (**C**) Aa, (**D**) the unloaded XG/PF-127 hydrogel, (**E**) ATMH, and (**F**) the loaded XG/PF-127 hydrogel.

2.2. TGA

TGA was performed for XG, PF-127, and prepared hydrogels as illustrated in Figure 2. The TGA curve of XG indicated a weight loss of 8 and 42% at 218 and 400 °C, respectively. The initial loss of weight was correlated with the release of volatile matter and absorbed water. On the other hand, stability can be seen by the TGA of PF-127 till 378 °C. Further rise in temperature resulted in the decomposition of PF127. A 95% weight loss was observed as temperature reached 423 °C. Increasing temperature led to a further decomposition of PF127, which still continued to entire degradation. The loss in the weight of PF-127 was due to the elimination of functional groups with the increase in temperature [19,20]. A degradation of 37% in weight was observed by TGA of XG/PF-127 as temperature approached 310 °C. This might be due to the removal of absorbed and bonded moisture and loss of hydroxyl groups. Similarly, a weight loss of 48% was seen at 487 °C. Further decomposition of developed hydrogel was seen with the increase in temperature. Comparing the thermal stability of XG and PF127 with the formulated hydrogel, we can demonstrate that the prepared hydrogel exhibited higher thermal stability than pure polymers, which basically indicated enhancement in the thermal stability of excipients after crosslinking among them [21].

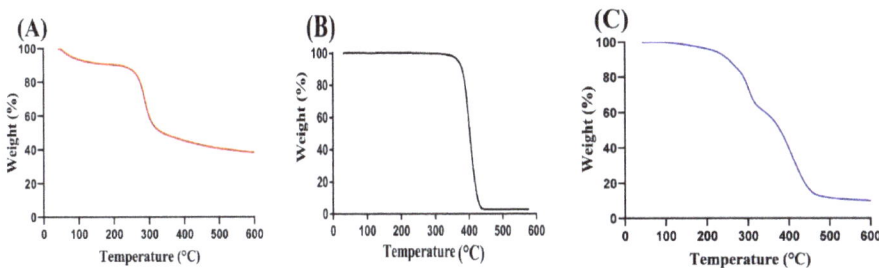

Figure 2. TGA of (**A**) XG, (**B**) PF-127, and (**C**) XG/PF-127 hydrogel.

2.3. DSC

The DSC curve of individual excipients and prepared hydrogel is shown in Figure 3. An exothermic peak was demonstrated at 65 °C while a broad endothermic peak was detected at 280 °C by XG's DSC. The endothermic and exothermic peaks of XG indicated moisture loss and thermal decomposition. The DSC of PF-127 exhibited a strong endothermic peak at 70 °C, representing the devastation of crystalline network of the PF-127 chain. Similarly, an exothermic peak was seen at 113 °C. The developed hydrogel and individual components are quite compatible, as seen by the prepared formulation's peaks migrating towards a higher glass transition temperature than the parent components [22] because of the higher intermolecular hydrogen bonding [23]. Therefore, greater thermal stability was indicated by the formulated hydrogel [24].

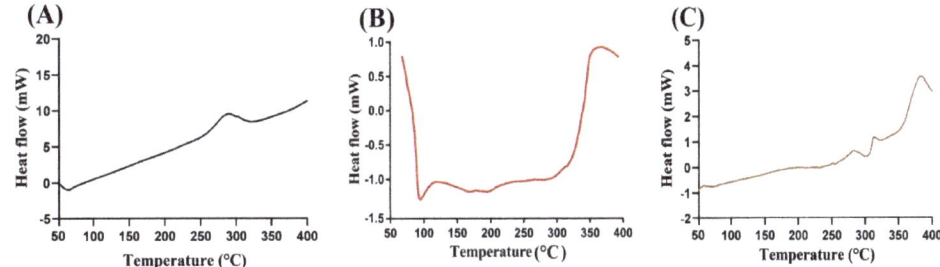

Figure 3. DSC of (**A**) XG, (**B**) PF-127, and (**C**) XG/PF-127 hydrogel.

2.4. XRD Analysis

The recorded XRD spectra of XG, PF127, and the formulated hydrogels are indicated in Figure 4. Prominent peaks were demonstrated by the XRD spectra of XG at $2\theta = 20.19°$, $32.10°$, and $48.40°$. XG is amorphous by nature and forms aggregates with the other side chains, hence restricting the proper packaging of polymer chains. Similarly, high sharp peaks of PF-127 were seen at $2\theta = 19.20°$, $21.83°$, and $27.10°$, respectively. The sharp high intense peaks of the PF-127 basically indicated its high stability and crystallinity. However, sharp and prominent peaks of XG and PF127 were replaced by dense peaks after crosslinking and polymerization reaction as indicated by the XRD analysis of the prepared hydrogel. All these factors indicate a decrease in pure reagent's crystallinity after polymerization. The reduction in crystallinity of the reagents as indicated by polymeric hydrogel may be the feature of the formation of the conjugate of Aa with XG and PF127 in the presence of MBA, hence indicating enhancement in the fraction of the amorphous phase [25,26].

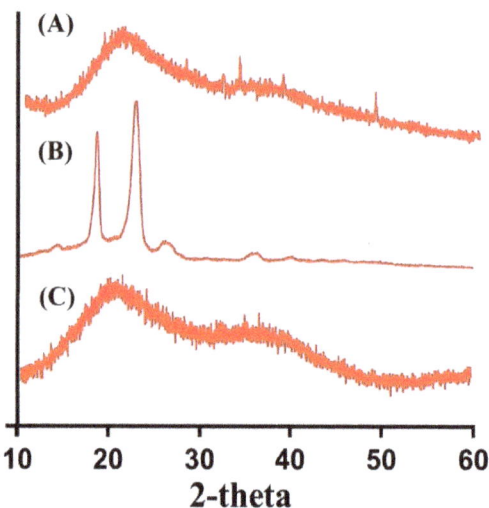

Figure 4. XRD of (**A**) XG, (**B**) PF-127, and (**C**) XG/PF-127 hydrogel.

2.5. SEM

The SEM of the XG/PF-127 hydrogel is illustrated in Figure 5. The prepared network displayed a hard surface with a few big pores, which may be connected to high crosslinking of the XG and PF-127 with the Aa content. The strong crosslinking among hydrogel reagents improved the mechanical strength and stability of the hydrogel; thus, it can be used as a controlled drug delivery carrier [27].

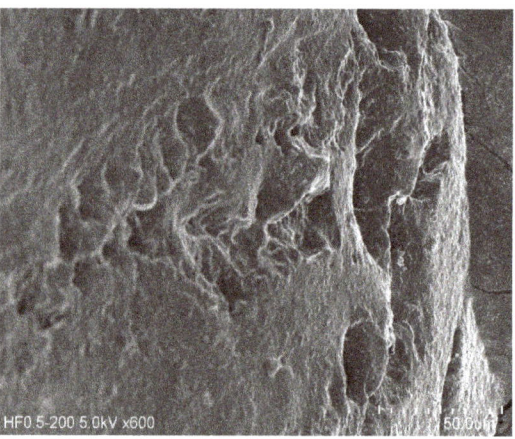

Figure 5. SEM of XG/PF-127 hydrogel.

2.6. Sol–Gel Analysis

The cross-linked and uncross-linked fractions of the synthesized hydrogel were estimated by sol–gel analysis (Table 1). Gel is the cross-linked while sol is the uncross-linked fraction of the formulated hydrogel. Both sol and gel fractions were influenced highly by the incorporated reagents of hydrogel. With an increase in the XG and PF127 feed ratios, the gel fraction rose. During polymerization reaction, free radicals are produced, which leads to the crosslinking of XG and PF-127 with Aa by MBA. Thus, as the feed ratios of XG and PF127 increase, more free radicals are generated in the same way. Thus, greater reactive sites are available for the monomer to crosslink with the polymers. Similarly, high incorporated feed ratios of Aa also resulted in high levels of gel formation. Aa plays a key role in polymerization of hydrogel reagents. Crosslinking density of the hydrogel is improved with the high incorporated contents of Aa, thus high gel fraction is achieved. In other words, we can say that like XG and PF127, high feed rations of Aa also result in high gel fraction [27]. Khalid and coworkers reported high gel fraction with the high feed ratios of hydrogel reagents [28]. Unlikely, a reduction was observed in the sol fraction with the high incorporated hydrogel contents [29]. Nasir et al. (2019) reported low sol with high gel fractions for the developed gels with their high incorporated contents [30].

Table 1. Drug loading and polymer volume fraction of XG/PF-127 hydrogels.

Formulation Code	Sol Fraction (%)	Gel Fraction (%)	Drug Loaded (mg)/350 mg of Dry Gel	
			Weight Method	Extraction Method
PXF-1	11	89	132.10 ± 0.71	130.84 ± 1.03
PXF-2	9	91	155.23 ± 0.93	152.42 ± 0.64
PXF-3	7	93	164.01 ± 1.10	163.02 ± 0.85
PXF-4	15	85	102.34 ± 0.78	101.04 ± 0.94
PXF-5	12	88	93.61 ± 0.93	91.23 ± 1.01
PXF-6	10	90	86.03 ± 1.03	84.87 ± 0.84
PXF-7	16	84	143.44 ± 0.87	142.31 ± 0.92
PXF-8	14	86	161.05 ± 0.91	159.92 ± 0.48
PXF-9	13	87	170.72 ± 0.83	168.63 ± 1.20

2.7. Porosity

The swelling and drug loading of the hydrogel and its sub-micro/nano particulate systems are dependent completely on their porosity. High porosity resulted in maximum swelling and loading of the drug. Porosity study was performed for the hydrogel formulations. Porosity was affected by different incorporated feed ratios of the hydrogel contents (Figure 6). With XG's increased feed ratios, a rise in porosity was seen. Similarly, an increase was seen in porosity with the high feed ratios of Aa. This may be correlated with the formation of highly viscous mixture during the polymerization process, which restricted the evaporation of bubbles; as a result, interconnected channels were produced. Water molecules penetrated into the hydrogel networks through these channels, and thus high porosity was achieved. Unlike in the case with XG and Aa, PF127 also affected the porosity, but in a reverse way. An increase in PF127 contents resulted in a low porosity due to the formation of a highly cross-linked network, which increased the hardness and decreased the pore size of the prepared network. Hence, it can be demonstrated that an increase in feed rations of PF127 and a drop in porosity are observed while an increase in porosity is seen with the high incorporated feed ratios of XG and Aa [31,32].

2.8. Swelling Study

The swelling degree of hydrogels is shown in Figure 7A, indicating a low swelling index at pH 1.2 and 4.6; however, it augmented considerably with the increase in pH of the medium, i.e., to pH 7.4. This change in swollen hydrogels due to pH changes can occur due to the development of osmotic swelling forces. These forces are generated by the carboxyl groups (Aa) present in the hydrogel network. Carboxyl groups begin to ionize at pH 4 and completely ionize above pH 6. The existence of more ionic groups in the hydrogel network at high pH resulted in maximum swelling. The prepared XG/PF-127 hydrogel consist of both COO^- and –COOH groups. These groups can change into one another under favorable conditions. At pH 1.2, –COO groups protonated into –COOH, but at basic pH 7.4, –COOH entirely deprotonated back into –COO groups. Moreover, XG contains O-acetyl, pyruvyl and unreacted hydroxyl groups, which also undergo deprotonation at pH > 6. Thus, high charge density was produced, and thus high electrostatic repulsion between negatively charged $–COO^-$ groups occurred, which led to high swelling of prepared hydrogel at higher pH values [33]. The hydrogen-bonding strength among hydrogel networks is most likely to be strengthened by the large ratio of protonated COOH groups, and thus low swelling was observed at pH 1.2 and 4.6 [30]. Similarly, hydrogel contents also have a significant impact on the hydrogel's swelling. A rise in swelling was seen with the high feed ratios of XG and Aa because of their high functional groups, while in case of PF-127, a drop was observed in swelling with the high levels of the incorporated PF-127 contents. The reason may be attributed to the formation of a hard and bulk network of hydrogel which did not allow the sufficient water molecules to enter into the hydrogel networks [34,35].

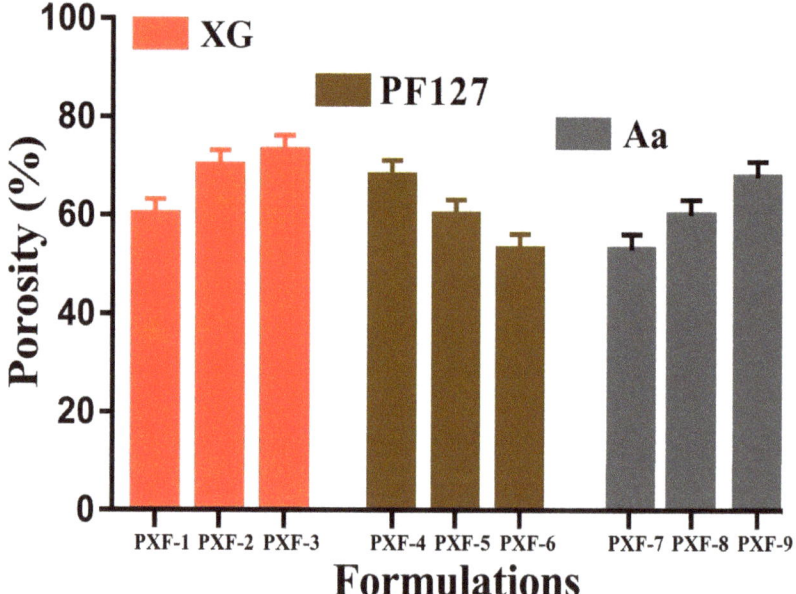

Figure 6. Effect of XG, PF-127, and Aa on porosity of XG/PF-127 hydrogel.

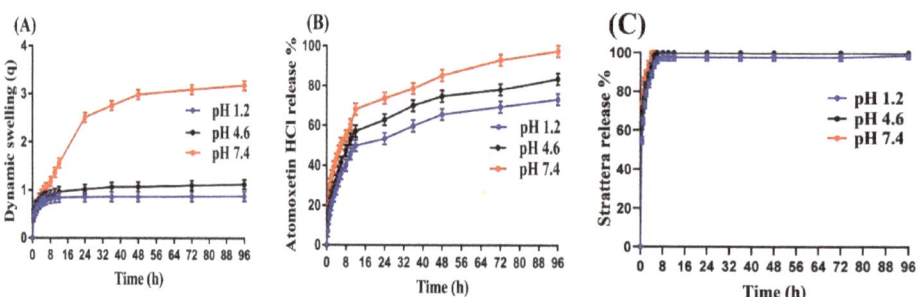

Figure 7. Effect of pH on (A) swelling, (B) drug release from XG/PF-127 hydrogel, and (C) commercial product Strattera.

2.9. Drug Loading and In Vitro Drug Release Studies

Drug loading was conducted for XG/PF-127 hydrogels as illustrated in Table 2. Like porosity and swelling, the contents of the hydrogel also affected the drug loading. Greater drug loading was detected with the high feed ratios of XG and Aa. The possible reason is the high porosity and swelling index of the hydrogel which occurred with the increase in

XG and Aa contents [36,37]. Contrary to XG and Aa, a drop was seen in drug loading by the synthesized matrix as the feed ratios of PF-127 were enhanced [38].

Table 2. Kinetic modeling release of drug from XG/PF-127 hydrogels.

F. Code	Zero Order r^2	First Order r^2	Higuchi r^2	Korsmeyer–Peppas r^2	N
PXF-1	0.9432	0.9920	0.8937	0.9582	0.5331
PXF-2	0.9356	0.9978	0.9220	0.9819	0.5562
PXF-3	0.9682	0.9870	0.9712	0.9754	0.5040
PXF-4	0.9548	0.9764	0.9692	0.9625	0.5128
PXF-5	0.9793	0.9954	0.9830	0.9922	0.5673
PXF-6	0.9450	0.9788	0.9706	0.9414	0.5468
PXF-7	0.9890	0.9903	0.9790	0.9378	0.5219
PXF-8	0.9063	0.9661	0.9332	0.9627	0.5493
PXF-9	0.9274	0.9845	0.9568	0.9439	0.5790

Drug release tests were carried out at pH 1.2, 4.6, and 7.4 for the prepared hydrogels and commercial product Strattera (Figure 7B,C). Maximum release of the drug was seen at high pH values due to the deprotonation of XG and Aa functional groups, whereas minimum drug release was detected at pH 1.2 due to the protonation of such functional groups [39,40]. The release studies of the Strattera indicated a rapid release of drug at all three pH values. Almost a 90% concentration of the drug was released at pH 7.4 within the initial 2 h, while in the case of pH 4.6 and 1.2, a concentration of more than 80% of the drug was released within the initial 2 h. Comparing the drug release of Strattera and that of the prepared hydrogel, we can predict that the drug was sustained successfully for extended period of time by the developed hydrogels.

Similar to swelling, the hydrogel's contents also have an impact on drug release. High drug release was observed with the high incorporated XG and Aa contents [41], while a decline was perceived in drug release with the high PF-127 contents [42] and vice versa.

2.10. Release Mechanism

To establish the sequence and drug release from the prepared networks, multiple kinetic models were computed using the release data of the synthesized hydrogels. "r" values determine the most suitable kinetic order. Table 2 indicates that the "r" values for the first order were higher than the "r" values of all other kinetic models. As a result, we can state that all formulations of the prepared hydrogel follow the first order of kinetics. The diffusion type is determined by the "n" value. If $0.45 > n$, Fickian diffusion occurs; otherwise, non-Fickian or anomalous transport similar to linked diffusion/polymer relaxation occurs. [43]. All formulations have "n" values between 0.5040 and 0.5790, indicating non-Fickian diffusion [44].

3. Conclusions

The free radical polymerization technique was adopted for the development of XG/PF-127 hydrogels. A stimuli-responsive effect was manifested by the graft copolymer. The controlled release of the ATMH was detected at pH 7.4 for 96 h. The compatibility among the various formulation components of the developed hydrogel was shown by FTIR. DSC and TGA showed an increase in the thermal stability of XG and PF-127 after the polymerization process. Similarly, SEM indicated a hard surface with few large pores through which water penetration occurred. After a polymerization process, XRD showed a reduction in the crystallinity of polymers. High swelling index was seen at pH 7.4, whereas at pH 1.2 and 4.6, low swelling was observed. Hence, it could be inferred from the results that the XG/PF-127 hydrogel can be employed as a potential agent for controlled drug delivery, which would be helpful not only in minimizing undesired GIT effects, but also in enhancing patient compliance and therapeutic output.

4. Materials and Methods

4.1. Materials

ATMH was obtained from Hetero labs limited (UNITE I), Telangana, India. Similarly, Xanthan gum was purchased from Tokyo Chemical Industry Co., Ltd., (Tokyo, Japan). Acrylic acid was purchased from Acros (Carlsbad, CA, USA). Likewise, pluronic F127 and ammonium persulfate were obtained from Sigma-Aldrich (Chemie GmbH, Riedstir-Steinheim, USA) and Showa (Tokyo, Japan). N,N'-Methylene bisacrylamide (MBA) was acquired from Alfa Aesar (Lancashire, UK), respectively.

4.2. Methods of Preparation

Xanthan gum (XG)- and pluronic (PF-127) [XG/PF-127]- based hydrogels were prepared by the free radical polymerization technique. A pre-weighed quantity of XG was dissolved in distilled water at 40 °C for 6 h. The PF-127 solution was mixed with the XG solution under constant stirring. Ammonium persulfate (APS) was poured into the mixture. After that, acrylic acid (Aa) was added in to the XG, PF127 and APS mixture. Finally, an MBA solution was mixed with the aforementioned mixture to crosslink the polymers and monomer on their specific sites. The entire mixture was kept on stirring until a transparent solution was formed, purged by nitrogen gas and then transferred into the glass molds, which were positioned in the water bath. The temperature of the water bath was maintained at 55 °C initially for 2 h, which was increased up to 65 °C later. The formulated gel was cut into 8 mm discs. A mixture of distilled water and ethanol was used for washing the prepared discs of gels. After that, hydrogel discs were placed for one week in a vacuum oven for dehydration. The dried discs were processed further for different studies. A set of formulations is shown in Table 3.

Table 3. Feed ratio scheme for formulation of XG/PF-127 hydrogels.

F. Code	Polymer (XG) g/30 g	Polymer (PF-127) g/30 g	Monomer (Aa) g/30 g	Initiator (APS) g/30 g	Cross-linker (MBA) g/30 g
PXF-1	0.080	0.200	4.0	0.1	0.2
PXF-2	0.120	0.200	4.0	0.1	0.2
PXF-3	0.160	0.200	4.0	0.1	0.2
PXF-4	0.050	0.300	4.0	0.1	0.2
PXF-5	0.050	0.350	4.0	0.1	0.2
PXF-6	0.050	0.400	4.0	0.1	0.2
PXF-7	0.050	0.200	4.5	0.1	0.2
PXF-8	0.050	0.200	5.0	0.1	0.2
PXF-9	0.050	0.200	5.5	0.1	0.2

4.3. Characterization

FTIR, TGA, DSC, XRD, and SEM were performed according to our previous publication [45].

4.4. Sol–Gel Anaylsis

The sol and gel contents of the synthesized hydrogel were estimated by sol–gel analysis. The unreacted part of the hydrogel is known as the sol fraction, while the reactant part is known as the gel fraction. The weighed discs of the prepared hydrogel were placed for 12 h in distilled water. After that, the discs were removed and positioned in the oven for dryness. The dehydrated discs were weighed again [46]. Equations (1) and (2) were applied for the estimation of sol and gel fractions.

$$\text{Sol fraction}\% = \frac{C_1 - C_2}{C_1} \times 100, \quad (1)$$

$$\text{Gel fraction} = 100 - \text{Sol fraction}. \quad (2)$$

Here, C_1 represents the initial weight of dried hydrogel before the extraction process, while C_2 is the final weight after the extraction process.

4.5. Porosity

The porosity study of the prepared hydrogels was performed in absolute ethanol. Hydrogel discs were taken, weighed (A_1) and immersed in ethanol for 3 days. After achieving equilibrium swelling, discs were removed and weighed (A_2) again [47]. Equation (3) was employed for the determination of porosity of the prepared hydrogels.

$$(\%) \text{ Porosity} = \frac{A_2 - A_1}{\rho V} \times 100. \tag{3}$$

Here, ρ is the density of absolute ethanol while V is the swelling volume of hydrogel discs.

4.6. Swelling Study

Weighed hydrogel discs were soaked in buffer solutions of pH 1.2, 4.6 and 7.4. The discs were removed at a specific interval of time, blotted with filter paper, weighed, and placed again in the respective media [48]. Dynamic swelling (q) was determined by Equation (4).

$$(q) = \frac{T_2}{T_1}, \tag{4}$$

where T_2 indicates the weight of the swollen hydrogel disc and T_1 represents the weight of the dried hydrogel disc at time t.

4.7. Drug Loading

A pH 7.4 phosphate buffer solution was used to prepare a 1% drug solution. For four days, precisely weighed hydrogel discs were submerged in the drug solution. Following that, discs were removed and cleaned with distilled water to remove any drugs that adhered to the hydrogel discs' surface. The cleaned discs were placed in a vacuum oven to dry [49].

4.8. In Vitro Drug Release Study

USP dissolution apparatus-II was employed for the drug release study. The release of the drug from the formulated hydrogel and commercial product Strattera (60 mg) (ELI LILLY and Co., Ltd.) was investigated at three different pH values, i.e., pH 1.2, 4.6 and pH 7.4. XG/PF127-loaded hydrogel discs and the commercial product were placed in 900 mL of each buffer solution with 50 rpm at 37 ± 0.5 °C. Samples of 5 mL were collected and fresh medium of the same quantity was added back to maintain constant sink conditions. The collected samples were analyzed by using a UV spectrophotometer (U-5100, 3 J2-0014, Tokyo, Japan) at the wavelength (λmax) of 226 nm [50].

4.9. Release Mechanism

The order and release mechanism of the drug from the prepared hydrogel was determined by using various kinetic models. The release data were fitted into zero-order, first-order, Higuchi, and Korsmeyer–Peppas models [51], respectively.

4.10. Statistical Analysis

SPSS Statistic software 22.0 (IBM Corp, Armonk, NY, USA) was performed for the statistical analysis of all experimental data. Differences between the tests were determined by Student's t-Test and considered significant statistically (p-value < 0.05).

Author Contributions: Conceptualization, P.-C.W.; data curation, M.S.; Y.-R.L., I.-H.C., A.K. and H.U.; formal analysis, M.S.; funding acquisition, P.-C.W.; investigation M.S.; methodology, P.-C.W. and N.S.A.-S.; project administration, I.-H.C.; supervision, P.-C.W.; writing—original draft, M.S.; and writing—review and editing, P.-C.W. All authors have read and agreed to the published version of the manuscript.

Funding: This research was funded by the National Science Council of Taiwan (MOST 110-2320-B-037-014-MY2 and NSTC 112-2320-B-037-014-MY3).

Conflicts of Interest: The authors declare no conflict of interest.

References

1. Ayano, G.; Yohannes, K.; Abraha, M. Epidemiology of attention-deficit/hyperactivity disorder (ADHD) in children and adolescents in Africa: A systematic review and meta-analysis. *Ann. Gen. Psychiatry* **2020**, *19*, 21. [CrossRef] [PubMed]
2. Williams, C.; Wright, B.; Partridge, I. Attention deficit hyperactivity disorder—A review. *Br. J. Gen. Pract.* **1999**, *49*, 563–571. [PubMed]
3. De Sousa, A.; Kalra, G. Drug therapy of attention deficit hyperactivity disorder: Current trends. *Mens Sana Monogr.* **2012**, *10*, 45. [CrossRef] [PubMed]
4. Drechsler, R.; Brem, S.; Brandeis, D.; Grünblatt, E.; Berger, G.; Walitza, S. ADHD: Current concepts and treatments in children and adolescents. *Neuropediatrics* **2020**, *51*, 315–335. [CrossRef]
5. Sevecke, K.; Battel, S.; Dittmann, R.; Lehmkuhl, G.; Döpfner, M. The effectiveness of atomoxetine in children, adolescents, and adults with ADHD. A systematic overview. *Der Nervenarzt* **2006**, *77*, 294, 297–300, 302. [CrossRef]
6. Teaima, M.H.; El-Nadi, M.T.; Hamed, R.R.; El-Nabarawi, M.A.; Abdelmonem, R. Lyophilized Nasal Inserts of Atomoxetine HCl Solid Lipid Nanoparticles for Brain Targeting as a Treatment of Attention-Deficit/Hyperactivity Disorder (ADHD): A Pharmacokinetics Study on Rats. *Pharmaceuticals* **2023**, *16*, 326. [CrossRef]
7. Mohanty, D.; Alsaidan, O.A.; Zafar, A.; Dodle, T.; Gupta, J.K.; Yasir, M.; Mohanty, A.; Khalid, M. Development of Atomoxetine-Loaded NLC In Situ Gel for Nose-to-Brain Delivery: Optimization, In Vitro, and Preclinical Evaluation. *Pharmaceutics* **2023**, *15*, 1985. [CrossRef] [PubMed]
8. Stanojević, G.; Medarević, D.; Adamov, I.; Pešić, N.; Kovačević, J.; Ibrić, S. Tailoring atomoxetine release rate from DLP 3D-printed tablets using artificial neural networks: Influence of tablet thickness and drug loading. *Molecules* **2020**, *26*, 111. [CrossRef]
9. Suhail, M.; Rosenholm, J.M.; Minhas, M.U.; Badshah, S.F.; Naeem, A.; Khan, K.U.; Fahad, M. Nanogels as drug-delivery systems: A comprehensive overview. *Ther. Deliv.* **2019**, *10*, 697–717. [CrossRef]
10. Khalid, I.; Ahmad, M.; Minhas, M.U.; Barkat, K. Synthesis and evaluation of chondroitin sulfate based hydrogels of loxoprofen with adjustable properties as controlled release carriers. *Carbohydr. Polym.* **2018**, *181*, 1169–1179. [CrossRef]
11. Malik, N.S.; Ahmad, M.; Alqahtani, M.S.; Mahmood, A.; Barkat, K.; Khan, M.T.; Tulain, U.R.; Rashid, A. β-cyclodextrin chitosan-based hydrogels with tunable pH-responsive properties for controlled release of acyclovir: Design, characterization, safety, and pharmacokinetic evaluation. *Drug Deliv.* **2021**, *28*, 1093–1108. [CrossRef]
12. Pawlicka, A.; Tavares, F.; Dörr, D.; Cholant, C.M.; Ely, F.; Santos, M.; Avellaneda, C. Dielectric behavior and FTIR studies of xanthan gum-based solid polymer electrolytes. *Electrochim. Acta* **2019**, *305*, 232–239. [CrossRef]
13. Malik, N.S.; Ahmad, M.; Minhas, M.U.; Murtaza, G.; Khalid, Q. Polysaccharide hydrogels for controlled release of acyclovir: Development, characterization and in vitro evaluation studies. *Polym. Bull.* **2017**, *74*, 4311–4328. [CrossRef]
14. Innocenzi, P.; Malfatti, L.; Piccinini, M.; Marcelli, A. Evaporation-induced crystallization of pluronic F127 studied in situ by time-resolved infrared spectroscopy. *J. Phys. Chem. A* **2010**, *114*, 304–308. [CrossRef] [PubMed]
15. Hu, H.; Yu, J.; Li, Y.; Zhao, J.; Dong, H. Engineering of a novel pluronic F127/graphene nanohybrid for pH responsive drug delivery. *J. Biomed. Mater. Res. Part A* **2012**, *100*, 141–148. [CrossRef] [PubMed]
16. Moharram, M.; Khafagi, M. Application of FTIR spectroscopy for structural characterization of ternary poly (acrylic acid)–metal–poly (vinyl pyrrolidone) complexes. *J. Appl. Polym. Sci.* **2007**, *105*, 1888–1893. [CrossRef]
17. Farheen, S.; Zubair, S.; Arpini, A.; Goud, G.; Soppari, S.; Kethavath, M. Formulation and Evaluation of Atomoxetine Hydrochloride Sustained Release Tablets. *PharmaTutor* **2017**, *5*, 76–92.
18. Patil, J.S.; Yadava, S.; Mokale, V.J.; Naik, J.B. Preparation and characterization of single pulse sustained release ketorolac nanoparticles to reduce their side-effects at gastrointestinal tract. In Proceedings of the International Conference on Advances in Chemical Engineering and Technology, Kollam, India, 16–18 October 2014; pp. 59–62.
19. Jin, J.; Mitome, T.; Egashira, Y.; Nishiyama, N. Phase control of ordered mesoporous carbon synthesized by a soft-templating method. *Colloids Surf. A Physicochem. Eng. Asp.* **2011**, *384*, 58–61. [CrossRef]
20. Tanaka, S.; Doi, A.; Nakatani, N.; Katayama, Y.; Miyake, Y. Synthesis of ordered mesoporous carbon films, powders, and fibers by direct triblock-copolymer-templating method using an ethanol/water system. *Carbon* **2009**, *47*, 2688–2698. [CrossRef]
21. Barkat, K.; Ahmad, M.; Usman Minhas, M.; Khalid, I.; Nasir, B. Development and characterization of pH-responsive polyethylene glycol-co-poly (methacrylic acid) polymeric network system for colon target delivery of oxaliplatin: Its acute oral toxicity study. *Adv. Polym. Technol.* **2018**, *37*, 1806–1822. [CrossRef]

22. Tummala, S.; Kumar, M.S.; Prakash, A. Formulation and characterization of 5-Fluorouracil enteric coated nanoparticles for sustained and localized release in treating colorectal cancer. *Saudi Pharm. J.* **2015**, *23*, 308–314. [CrossRef] [PubMed]
23. Gandhi, A.; Jana, S.; Sen, K.K. In-vitro release of acyclovir loaded Eudragit RLPO® nanoparticles for sustained drug delivery. *Int. J. Biol. Macromol.* **2014**, *67*, 478–482. [CrossRef] [PubMed]
24. Dey, P.; Maiti, S.; Sa, B. Gastrointestinal delivery of glipizide from carboxymethyl locust bean gum–Al3+–alginate hydrogel network: In vitro and in vivo performance. *J. Appl. Polym. Sci.* **2013**, *128*, 2063–2072. [CrossRef]
25. Lee, C.-T.; Huang, C.-P.; Lee, Y.-D. Synthesis and characterizations of amphiphilic poly (l-lactide)-grafted chondroitin sulfate copolymer and its application as drug carrier. *Biomol. Eng.* **2007**, *24*, 131–139. [CrossRef]
26. Chang, C.; Duan, B.; Zhang, L. Fabrication and characterization of novel macroporous cellulose–alginate hydrogels. *Polymer* **2009**, *50*, 5467–5473. [CrossRef]
27. Khanum, H.; Ullah, K.; Murtaza, G.; Khan, S.A. Fabrication and in vitro characterization of HPMC-g-poly (AMPS) hydrogels loaded with loxoprofen sodium. *Int. J. Biol. Macromol.* **2018**, *120*, 1624–1631. [CrossRef]
28. Khalid, I.; Ahmad, M.; Minhas, M.U.; Barkat, K. Preparation and characterization of alginate-PVA-based semi-IPN: Controlled release pH-responsive composites. *Polym. Bull.* **2018**, *75*, 1075–1099. [CrossRef]
29. Dergunov, S.A.; Nam, I.K.; Mun, G.A.; Nurkeeva, Z.S.; Shaikhutdinov, E.M. Radiation synthesis and characterization of stimuli-sensitive chitosan–polyvinyl pyrrolidone hydrogels. *Radiat. Phys. Chem.* **2005**, *72*, 619–623. [CrossRef]
30. Nasir, N.; Ahmad, M.; Minhas, M.U.; Barkat, K.; Khalid, M.F. pH-responsive smart gels of block copolymer [pluronic F127-co-poly (acrylic acid)] for controlled delivery of Ivabradine hydrochloride: Its toxicological evaluation. *J. Polym. Res.* **2019**, *26*, 212. [CrossRef]
31. Yin, L.; Fei, L.; Cui, F.; Tang, C.; Yin, C. Superporous hydrogels containing poly (acrylic acid-co-acrylamide)/O-carboxymethyl chitosan interpenetrating polymer networks. *Biomaterials* **2007**, *28*, 1258–1266. [CrossRef]
32. Ranjha, N.M.; Qureshi, U.F. Preparation and characterization of crosslinked acrylic acid/hydroxypropyl methyl cellulose hydrogels for drug delivery. *Int. J. Pharm. Pharm. Sci.* **2014**, *6*, 410.
33. Malik, N.S.; Ahmad, M.; Minhas, M.U.; Tulain, R.; Barkat, K.; Khalid, I.; Khalid, Q. Chitosan/xanthan gum based hydrogels as potential carrier for an antiviral drug: Fabrication, characterization, and safety evaluation. *Front. Chem.* **2020**, *8*, 50. [CrossRef]
34. Qudah, Y.; Raafat, A.; Ali, A. Removal of some heavy metals from their aqueous solutions using 2-Acrylamido-2-Methyl-1-propane sulfonic acid/polyvinyl alcohol copolymer hydrogels prepared by gamma irradiation. *Arab J. Nucl. Sci. Appl.* **2013**, *46*, 80–91.
35. Şanlı, O.; Ay, N.; Işıklan, N. Release characteristics of diclofenac sodium from poly (vinyl alcohol)/sodium alginate and poly (vinyl alcohol)-grafted-poly (acrylamide)/sodium alginate blend beads. *Eur. J. Pharm. Biopharm.* **2007**, *65*, 204–214. [CrossRef]
36. Murthy, P.K.; Mohan, Y.M.; Sreeramulu, J.; Raju, K.M. Semi-IPNs of starch and poly (acrylamide-co-sodium methacrylate): Preparation, swelling and diffusion characteristics evaluation. *React. Funct. Polym.* **2006**, *66*, 1482–1493. [CrossRef]
37. Sullad, A.G.; Manjeshwar, L.S.; Aminabhavi, T.M. Novel pH-sensitive hydrogels prepared from the blends of poly (vinyl alcohol) with acrylic acid-graft-guar gum matrixes for isoniazid delivery. *Ind. Eng. Chem. Res.* **2010**, *49*, 7323–7329. [CrossRef]
38. Suhail, M.; Hung, M.-C.; Chiu, I.-H.; Vu, Q.L.; Wu, P.-C. Preparation and in-vitro characterization of 5-aminosalicylic acid loaded hydrogels for colon specific delivery. *J. Mater. Res. Technol.* **2022**, *21*, 339–352. [CrossRef]
39. El-Hag Ali, A. Removal of heavy metals from model wastewater by using carboxymehyl cellulose/2-acrylamido-2-methyl propane sulfonic acid hydrogels. *J. Appl. Polym. Sci.* **2012**, *123*, 763–769. [CrossRef]
40. Al-Tabakha, M.M.; Khan, S.A.; Ashames, A.; Ullah, H.; Ullah, K.; Murtaza, G.; Hassan, N. Synthesis, Characterization and Safety Evaluation of Sericin-Based Hydrogels for Controlled Delivery of Acyclovir. *Pharmaceuticals* **2021**, *14*, 234. [CrossRef]
41. Paloma, M.; Enobakhare, Y.; Torrado, G.; Torrado, S. Release of amoxicillin from polyionic complexes of chitosan and poly (acrylic acid). Study of polymer/polymer and polymer/drug interactions within the network structure. *Biomaterials* **2003**, *24*, 1499–1506.
42. Akash, M.S.H.; Rehman, K.; Li, N.; Gao, J.-Q.; Sun, H.; Chen, S. Sustained delivery of IL-1Ra from pluronic F127-based thermosensitive gel prolongs its therapeutic potentials. *Pharm. Res.* **2012**, *29*, 3475–3485. [CrossRef]
43. Siepmann, J.; Peppas, N.A. Modeling of drug release from delivery systems based on hydroxypropyl methylcellulose (HPMC). *Adv. Drug Deliv. Rev.* **2012**, *64*, 163–174. [CrossRef]
44. Korsmeyer, R.; Gurny, R.; Doelker, E.; Buri, P.; Peppas, N. Mechanisms of potassium chloride release from compressed, hydrophilic, polymeric matrices: Effect of entrapped air. *J. Pharm. Sci.* **1983**, *72*, 1189–1191. [CrossRef] [PubMed]
45. Suhail, M.; Wu, P.-C.; Minhas, M.U. Development and characterization of pH-sensitive chondroitin sulfate-co-poly (acrylic acid) hydrogels for controlled release of diclofenac sodium. *J. Saudi Chem. Soc.* **2021**, *25*, 101212. [CrossRef]
46. Ullah, K.; Khan, S.A.; Murtaza, G.; Sohail, M.; Manan, A.; Afzal, A. Gelatin-based hydrogels as potential biomaterials for colonic delivery of oxaliplatin. *Int. J. Pharm.* **2019**, *556*, 236–245. [CrossRef]
47. Zia, M.A.; Sohail, M.; Minhas, M.U.; Sarfraz, R.M.; Khan, S.; de Matas, M.; Hussain, Z.; Abbasi, M.; Shah, S.A.; Kousar, M. HEMA based pH-sensitive semi IPN microgels for oral delivery; a rationale approach for ketoprofen. *Drug Dev. Ind. Pharm.* **2020**, *46*, 272–282. [CrossRef] [PubMed]
48. Ijaz, H.; Tulain, U.R.; Azam, F.; Qureshi, J. Thiolation of arabinoxylan and its application in the fabrication of pH-sensitive thiolated arabinoxylan grafted acrylic acid copolymer. *Drug Dev. Ind. Pharm.* **2019**, *45*, 754–766. [CrossRef] [PubMed]
49. Barkat, K.; Ahmad, M.; Minhas, M.U.; Khalid, I.; Malik, N.S. Chondroitin sulfate-based smart hydrogels for targeted delivery of oxaliplatin in colorectal cancer: Preparation, characterization and toxicity evaluation. *Polym. Bull.* **2020**, *77*, 6271–6297. [CrossRef]

50. Khan, S.; Ranjha, N.M. Effect of degree of cross-linking on swelling and on drug release of low viscous chitosan/poly (vinyl alcohol) hydrogels. *Polym. Bull.* **2014**, *71*, 2133–2158. [CrossRef]
51. Peppas, N.A.; Sahlin, J.J. A simple equation for the description of solute release. III. Coupling of diffusion and relaxation. *Int. J. Pharm.* **1989**, *57*, 169–172. [CrossRef]

Disclaimer/Publisher's Note: The statements, opinions and data contained in all publications are solely those of the individual author(s) and contributor(s) and not of MDPI and/or the editor(s). MDPI and/or the editor(s) disclaim responsibility for any injury to people or property resulting from any ideas, methods, instructions or products referred to in the content.

Article

Functional Hydrogels for Agricultural Application

Romana Kratochvílová [1], Milan Kráčalík [2], Marcela Smilková [1], Petr Sedláček [1], Miloslav Pekař [1], Elke Bradt [2], Jiří Smilek [1], Petra Závodská [1] and Martina Klučáková [1,*]

[1] Faculty of Chemistry, Brno University of Technology, Purkyňova 464/118, CZ-61200 Brno, Czech Republic; kolajova@fch.vutbr.cz (R.K.); smilkova@fch.vut.cz (M.S.); sedlacek-p@fch.vut.cz (P.S.); pekar@fch.vut.cz (M.P.); smilek@fch.vut.cz (J.S.); petra.zavodska@vut.cz (P.Z.)

[2] Institute of Polymer Science, Johannes Kepler University, Altenbergerstrasse 69, 4040 Linz, Austria; milan.kracalik@jku.at (M.K.); elke.bradt@jku.at (E.B.)

* Correspondence: klucakova@fch.vutbr.cz

Abstract: Ten different hydrogels were prepared and analyzed from the point of view of their use in soil. FT-IR spectra, morphology, swelling ability, and rheological properties were determined for their characterization and appraisal of their stability. The aim was to characterize prepared materials containing different amounts of NPK as mineral fertilizer, lignohumate as a source of organic carbon, and its combination. This study of stability was focused on utility properties in their application in soil—repeated drying/re-swelling cycles and possible freezing in winter. Lignohumate supported the water absorbency, while the addition of NPK caused a negative effect. Pore sizes decreased with NPK addition. Lignohumate incorporated into polymers resulted in a much miscellaneous structure, rich in different pores and voids of with a wide range of sizes. NPK fertilizer supported the elastic character of prepared materials, while the addition of lignohumate shifted their rheological behavior to more liquid. Both dynamic moduli decreased in time. The most stable samples appeared to contain only one fertilizer constituent (NPK or lignohumate). Repeated re-swelling resulted in an increase in elastic character, which was connected with the gradual release of fertilizers. A similar effect was observed with samples that were frozen and defrosted, except samples containing a higher amount of NPK without lignohumate. A positive effect of acrylamide on superabsorbent properties was not confirmed.

Keywords: superabsorbent; swelling; fertilizer; lignohumate; rheology; stability

Citation: Kratochvílová, R.; Kráčalík, M.; Smilková, M.; Sedláček, P.; Pekař, M.; Bradt, E.; Smilek, J.; Závodská, P.; Klučáková, M. Functional Hydrogels for Agricultural Application. *Gels* 2023, 9, 590. https://doi.org/10.3390/gels9070590

Academic Editors: Dong Zhang, Jintao Yang, Xiaoxia Le and Dianwen Song

Received: 19 June 2023
Revised: 19 July 2023
Accepted: 20 July 2023
Published: 22 July 2023

Copyright: © 2023 by the authors. Licensee MDPI, Basel, Switzerland. This article is an open access article distributed under the terms and conditions of the Creative Commons Attribution (CC BY) license (https:// creativecommons.org/licenses/by/ 4.0/).

1. Introduction

Superabsorbent polymers are hydrogels created by freely crosslinked three-dimensional networks of flexible polymeric chains homogenously distributed in a water dispersion medium. From a structural point of view, they are crosslinked polyelectrolytes [1]. They have the unique ability to absorb and retain relatively large amounts of water or water solutions in their structure as a result of rising entropy inside the polymer chain network [1–6]. Their field of application is very wide. They are suitable where the absorption of liquids is required. The most common application is in sanitary equipment. They are very often used as targeted carriers of special substances, e.g., in agriculture, to ensure sufficient humidity for cultivated crops [7–12].

Inspired by published works [2,3,7–12] and their results, the composite materials for the controlled release of mineral nutrients and humic substances for agricultural application were developed in our previous work [13]. In this work, we are focused on the characterization of prepared materials by means of FT-IR, SEM, and rheology. The aim of this work is to investigate materials from the point of view of their agricultural application. It means that the effect of freezing and repeated re-swelling on the mechanical properties of storage and loss moduli was investigated.

In the beginning, we investigated results available in studies published for superabsorbent polymers with similar compositions. Li et al. [8] synthetized a poly(acrylic acid)/sodium humate superabsorbent composite and studied its water absorbency properties. The effects of the initial concentration of acrylic acid, its neutralization degree, contents of crosslinker (N,N'-methylenebisacrylamide), initiator (ammonium persulfate), and sodium humate as an active filler were investigated. It was found that the water absorbency was much higher in distilled water compared with the NaCl solution. The addition of sodium humate supported material swelling up to 20% wt. A decrease in swelling was observed for higher additions. Water absorbency gradually decreased with the re-swelling cycles up to ~70% wt. of the initial value for the fifth cycle. Materials were characterized by FT-IR spectrometry and thermogravimetry. Similar studies were realized by Chu et al. [9,10]. They synthetized superabsorbents based on the same starting materials, but their achieved water absorbency was much lower [9]. They studied swelling in various saline solutions and observed a linear relationship between the saturated water absorbency and the minus square root of the ionic strength of the external medium. The water absorbency of PAA-AM/KHA in various salt solutions had the following order: $NH_4Cl_{(aq)} = KCl_{(aq)} = NaCl_{(aq)} > MgCl_{2(aq)} > CaCl_{2(aq)} > AlCl_{3(aq)} > FeCl_{3(aq)}$ [10]. Liu et al. [11] prepared a multifunctional superabsorbent based on chitosan, g-poly(acrylic acid), and sodium humate. They stated that sodium humate enhanced water absorbency, and the content of 10 wt% sodium humate gave the best absorption. Gao et al. [12] investigated the effects of N, reaction temperature, and contents of initial materials on water absorption. They confirmed the positive effect of the addition of humic acid on water absorbency.

Smart hydrogels for agricultural applications were recently developed and investigated [14,15]. Obtained results and characteristics were described in several reviews [16–19]. Zhang et al. [14] developed new multifunctional smart soils with double-layer structures consisting of zwitterionic, thermo-responsive poly(NIPAM-co-VPES), and poly(NIPAM-co-SBAA) aerogels doped in the upper layer and the bottom bed of the soils, respectively. Pushpamalar et al. [15] developed eco-friendly smart hydrogels for soil conditioning and sustain release fertilizer generated from biomass waste. A review focused on cellulose-based hydrogel materials and their prospective applications in agricultural activity was published by Kabir et al. [16]. Another review encompassed the latest developments in the field of smart hydrogel synthesis based on their unique features and different aspects of their responsive behaviors was published by Sikdar et al. [17]. Azeem et al. [18] published the review that spotlighted application prospects of three-dimensional hydrogel in agriculture. Hydrogel functioning, significance, advantages, mechanism of fertilizer release, and agriculture-specific applications were comprehensively described. In the review of Chakraborty et al. [19], the use of different nanocomposite materials developed for nutrient management in agriculture was summarized with a major focus on their synthesis and characterization techniques and application aspects in plant nutrition, along with addressing constraints and future opportunities of this domain.

Effects of structural variables on water absorbency and rheological behavior of superabsorbents based on acrylic polymers were investigated by Ramazani-Harandi et al. [20]. Their results showed a linear correlation between water absorbency and storage modulus over the rubber-elastic plateau. The swelling improvement and swollen gel strength were observed with increased crosslinker concentration. Zaharia et al. [21] enriched a poly(acrylic acid-co-N,N'-methylene-bis-acrylamide) composite hydrogel with bacterial cellulose. Infrared spectra, thermogravimetric curve, XRD diffraction pattern, SEM micrograph, swelling degree, and rheological properties were determined and investigated for both basic polymer and enriched material. These results can be used for the comparison with our experimental data. A review focused on the rheological properties and behavior of polymer hydrogels for different industrial applications was published by Rehman and Shah [22].

NPK, as a mineral nutrient used in agriculture, is often combined with different organic manures in order to achieve suitable fertilization for soil health, plant growth, and crop yields [23–25]. In this work, the amendment of manure was replaced by lignohumate as the source of organic carbon. It is an industrially produced analog of natural humic substances produced by the thermal processing of technical lignosulfonate, which is based on the oxidation and hydrolytic destruction of lignin-containing raw materials [26–29]. Due to its smaller molecular size and weight in comparison to humic substances isolated from native sources, it is more soluble and has a similar character to the most active fractions of humic substances dissolved in soil solution [30–33]. The characterization of lignohumate using elemental analysis and spectral methods and its comparison with humic substances isolated from different raw materials can be found in refs. [27,34]. The unique properties and supramolecular character of lignohumate can result in the promotion of plant growth connected with its utilization for agricultural and horticultural purposes [27,35–40].

The above-mentioned references provide characteristics for the comparison with functional materials investigated in this study. Our materials are based on poly(acrylic acid) (AA). N,N′-methylenebisacrylamide (MBA) was used as a crosslinker, and potassium peroxydisulfate (KPS) as an initiator. The polymer was enriched by potassium lignohumate (LH) as a source of organic carbon and NPK as a mineral nutrient [13]. The effect of their additions on the properties of prepared functional materials was investigated. In this work, the addition of acrylamide (AM), which was used in several references [2,3,5,10], was also studied in order to consider its influence on the properties of superabsorbent materials. However, the use of acrylamide is problematic because of its toxicity [41–44]; therefore, the superabsorbent polymers containing higher amounts of acrylamide are not friendly for agricultural use.

2. Results and Discussion

Ten different superabsorbent materials were prepared in order to investigate the effect of inorganic and organic nutrients, as well as acrylamide, on the material properties. The composition of prepared materials is described in Table 1. AA, KOH, MBA, and KPS were used for the preparation of all superabsorbent materials in the same amounts; therefore, only constituents differed in individual samples are listed in Table 1. Samples A...F contained mineral nutrient NPK in lower (A, C, E) and higher (B, D, F) amounts. Samples C...F combined NPK with LH, and samples G and H contained only LH. Samples I and J were without inorganic and organic nutrients and were analyzed for comparison with others. Alternatively, materials containing acrylamide (E, F, G, J) were prepared for the comparison of their properties and behavior. In consideration of AM toxicity, its content was relatively low. The AM/AA ratio was 5/95 in comparison with [45], where the ratios were from 30/70 to 70/30.

Table 1. The composition of prepared superabsorbent materials.

Sample	A	B	C	D	E	F	G	H	I	J
AM (g)	0	0	0	0	0.75	0.75	0.75	0	0	0.75
NPK (g)	0.660	6.602	0.660	6.602	0.660	6.602	0	0	0	0
LH (g)	0	0	1	1	1	1	1	1	0	0

2.1. FT-IR Spectrometry

FT-IR spectra of prepared materials, NPK, and lignohumate were collected. Samples were measured in dry forms by means of the ATR technique. In Figure 1, spectra for NPK, LH, and sample I are compared. Spectra were collected for the comparison of different prepared superabsorbent polymers and the identification of their constituents. The spectrum of NPK was described in detail in [45–47]. Peaks between 3000 and 3500 cm^{-1} can be assigned to stretching modes of O-H and N-H bonds in urea, potassium dihydrogen phosphate, and ammonium dihydrogen phosphate. The shoulder around 2800 cm^{-1} corresponds to the O-H stretching. Peaks between 1300 and 1400 cm^{-1} belong to P=O stretching, 1160 cm^{-1}

is related to P-OH stretching, and 1070 cm^{-1} is attributed to the HO-P-OH bending. The O-N=P and O-N bonds in ammonium dihydrogen phosphate appear as peaks between 700 and 800 cm^{-1} [45,46]. The spectrum of lignohumate was analyzed and compared with natural humic substances in ref. [29]. The broad peak above 3000 cm^{-1} belongs to O-H stretching and N–H stretching in different functional groups and overlaps vibrations of C-H groups in aromatic structures. Bands between 2800 and 3000 cm^{-1} can be assigned to C-H and CH_2 groups. Aromatic C=C vibrations and C=O stretching of quinone and amide groups can be observed between 1600 and 1640 cm^{-1}. N–H deformation and C=N stretching of amides appear between 1500 and 1520 cm^{-1}. Bands between 1400 and 1500 cm^{-1} can be assigned to C-H bending of the CH_3 group, O-H deformation, and C-O stretching phenolic groups. C=O stretching of aryl esters and C-O stretching of aryl ethers and phenols can be seen between 1200 and 1300 cm^{-1}. C-O stretching of secondary alcohols, ethers, and polysaccharides or polysaccharide-like substances appear between 1000 and 1200 cm^{-1} [34,48]. Superabsorbent polymers based on AA and AM, as well as the starting materials, were characterized in ref. [49]. Peaks between 3000 and 3500 cm^{-1} belong to O-H and N-H stretching. The peak of C-H in the methylene group is visible at 2940 cm^{-1}. Two bands between 1550 and 1720 cm^{-1} can be assigned to C=C and C-O vibrations. C-O vibrations also appear between 1100 and 1200 cm^{-1}.

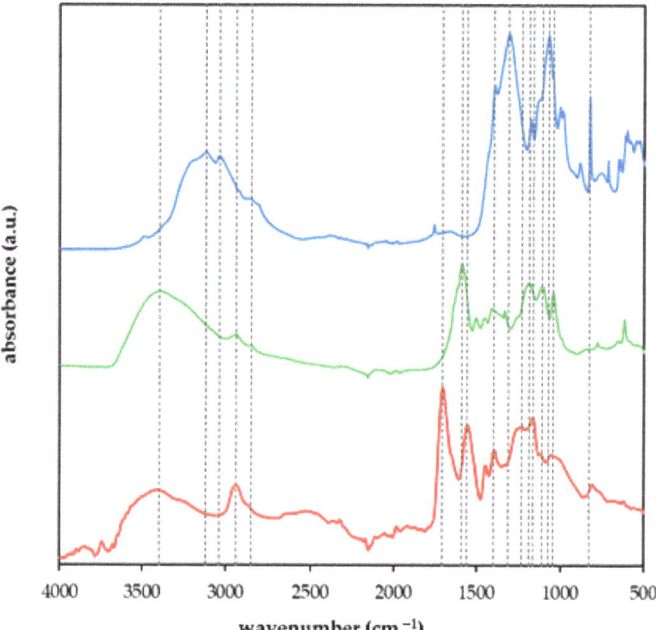

Figure 1. FT-IR spectra of NPK (blue), lignohumate (green), and sample A (red). Dotted lines are added for easy comparison of peaks in spectra.

FT-IR spectra of some prepared hydrogels are shown in Figure 2. We can see that the presence of NPK fertilizer came through mainly in the spectrum of sample D which contains ten times higher amounts in comparison with sample C. Characteristic peaks observable at 1390 cm^{-1}, 1140 cm^{-1}, 1160 cm^{-1}, and 1100 cm^{-1} are related to P=O stretching, P-OH stretching, and 1070 cm^{-1}, and HO-P-OH bending, respectively. The peaks shifted slightly in comparison with the spectrum of NPK (Figure 1). Samples G and H are without NPK and contain LH as a source of organic carbon. The difference between the two samples is the addition of AM (sample G). No noticeable differences in the spectra of both samples were observed.

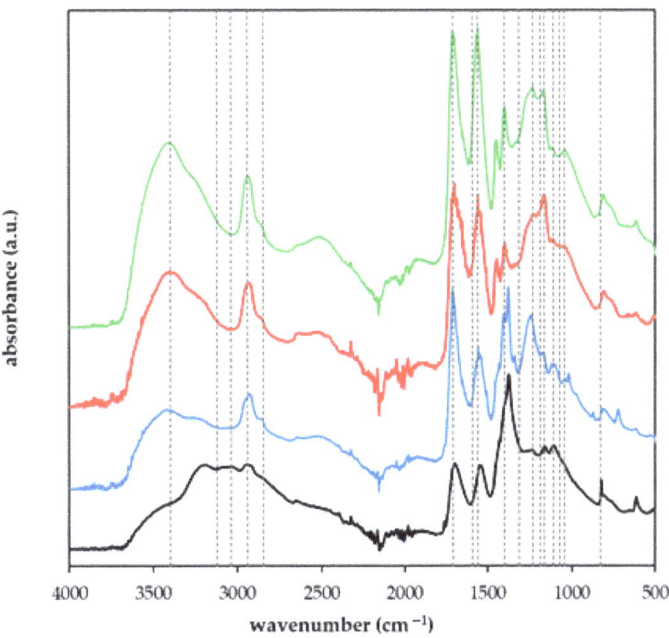

Figure 2. FT-IR spectra of samples C (blue), D (black), G (red), and H (green). Dotted lines are added for easy comparison of peaks in spectra.

2.2. SEM Analysis

The morphology of superabsorbent polymers was characterized by means of scanning electron microscopy (SEM). At first, the prepared samples before swelling were analyzed in order to discover possible changes in their surface given by different constituents added to the basic polymer constituents. In Figure 3, the photos of samples with low and high contents of NPK and the effect of LH addition are shown.

Figure 3. SEM photos of samples A with low content of NPK (**a**), B with high content of NPK (**b**), and D with high content of NPK and LH (**c**) before swelling.

As can be seen, the surface of sample A has a wrinkly character without visible pores and voids (Figure 3a). The surface morphology changed if the content of NPK increased (ten times). The crystals of NPK constituents (urea, potassium dihydrogen phosphate, and ammonium dihydrogen phosphate) are visible on the surface, which became more heterogeneous (Figure 3b). A similar effect on the surface morphology of superabsorbent polymers (if inorganic nutrients were incorporated in them) was observed in refs. [45,50,51].

The deposition of NPK fertilizer and the crystals of their constituents deposited onto pores walls were also observed in the swollen state in the higher magnification [45,51]. The surface roughness also increased after LH addition (Figure 3c) which can suppose the swelling ability of the superabsorbent (see Section 2.3.). The positive effect of humic substances on the formation of the porous structure was also observed in refs. [9,12,52].

The surface morphology of superabsorbent polymers in lower magnification was not affected by the addition of NPK, LH, and AM (Figure 4). Prepared samples (before swelling) had similar characteristics. The surface heterogeneity caused by the increased amount of NPK incorporated into polymers was, in the lower magnitude, less visible (not shown). The addition of LH resulted in the brown coloration of the prepared samples. Although some authors (e.g., [53]) observed the increase in hydrogel porosity in the case of higher content of AM, the low AM amount added in polymers had no similar effect. In Figure 5, the comparison of swollen and freeze-dried hydrogels is shown. It was observed that the NPK addition resulted in a decrease in pore size. While the sizes of pores and voids in the hydrogel with low content of NPK were several tens of μm (Figure 5a), it decreased in magnitude if the addition of NPK was higher (Figure 5b). The surface of pores was covered by NPK fertilizer, similarly as in refs. [45,51]. The incorporation of LH in the superabsorbent polymer changed the structure of the swollen hydrogel. We can see that it is much miscellaneous and rich in different pores and voids of a wide range of their sizes. The detail of the structure formed by the combination of NPK and LH is shown in Figure 6a. This structure was detected only in the case of sample C; the higher content of NPK resulted in the collapse of this miscellaneous character, and the crystals originating from NPK covered the pore surface. Although the addition of AM was not too high in comparison with other studies [2,3,5,10,49,54–56], its presence in hydrogel significantly changed the shape of pores and voids from spherical to oblong slits. This observation can influence the swelling behavior of samples enriched with AM, as described in Section 2.3.

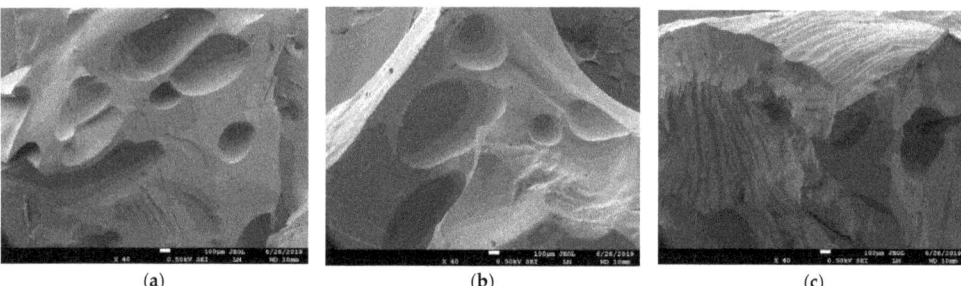

Figure 4. SEM photos of samples C with low content of NPK and LH (**a**), G with AM and LH (**b**), and H with LH (**c**) before swelling.

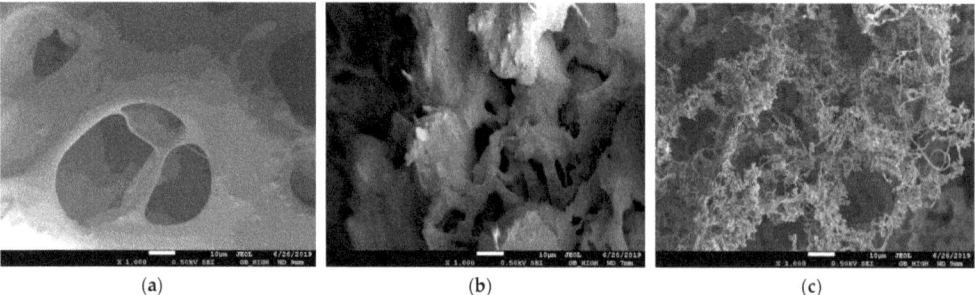

Figure 5. SEM photos of samples A with low content of NPK (**a**), B with high content of NPK (**b**), and C with low content of NPK and LH (**c**) after swelling and freeze-drying.

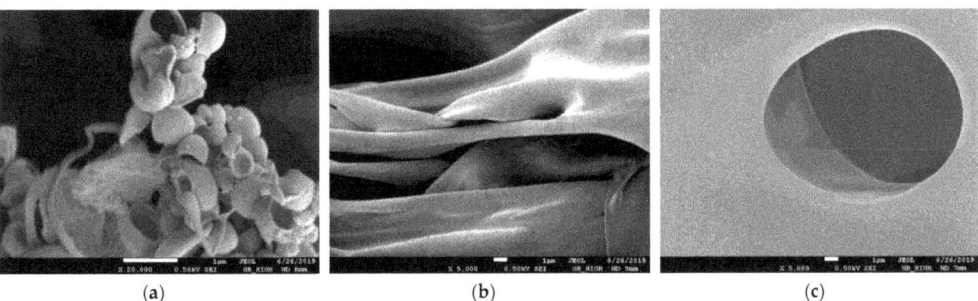

(a) (b) (c)

Figure 6. SEM photos of samples C with low content of NPK and LH in detail (**a**), G with AM and LH (**b**), and H with LH (**c**) after swelling and freeze-drying.

2.3. Water Absorbency

The swelling degrees of superabsorbent polymers in deionized and tap water are shown in Figure 7. The values were determined on the basis of weight increase after 24 h. The swelling kinetics was studied in our previous work [13], and it was found that the swelling was relatively fast, and the equilibrium was achieved over several hours. Similar findings were published in refs. [3,5,10]. Dadhanyia et al. [5] observed maximum swelling degree after 20 h. Some superabsorbent polymers prepared by Raju et al. [3] achieved the maximum swelling degrees instantly, others in several minutes [57,58]. As can be seen, all prepared samples exhibited very good swelling properties. The samples differed from each other by their special compositions, as shown in Table 1. Higher contents of NPK in the structure of the sample induced a significant negative effect on swelling properties. The lowest values were obtained for samples containing higher amounts of NPK which caused the depression of swelling. Samples B, D, and F exhibited a much lower ability to absorb surrounding water. In contrast, samples with the addition of lignohumate swelled significantly better than samples without lignohumate. The addition of LH supported the absorption of water, which resulted in the greatest degree of swelling of samples G and H (without NPK) and an increase in swelling degree for samples containing a combination of NPK and LH (C, D, E, and F) in comparison with samples enriched by NPK alone LH (A and B). An adequate number of humic substances in the hydrogel polymeric network enhanced the hydrophilicity of superabsorbent samples. Free functional groups of humate (-OH, -COOH, -NH_2, -SO_3, quinonyl groups) interacted with acylamino and carboxyl groups of superabsorbent polymers and caused a collaborative absorbent effect. The positive effect of humic substances on swelling was observed in recent works [8,10,12,59,60].

Superabsorbents containing potassium or sodium humate [8,10,12] evinced both a higher swelling rate and a greater amount of absorbed water. However, some authors noted that the efficiency of humic addition to enhance swelling reached a maximum value that the addition of higher amounts of humic substances had no additional influence on swelling [12] or, conversely, began to have a negative influence [8–11]. A recommended allowance of humic substances differed between 3 and 30% wt. Chu et al. [9] showed that with the addition of a superabsorbent (1% wt.), water retention in soil increased 31 times and 40 times with the addition of a superabsorbent enriched with sodium humate. The increase in moisture in soil enriched by superabsorbent polymers was observed in recent work [61]. Our content of LH was in the lower limit. The increase in ionic strength usually suppressed the ability of the hydrogel to absorb water, and their degrees of swelling decreased in the presence of salts [10,11]. It corresponds with the increase in swelling degree in tap water observed for all prepared samples.

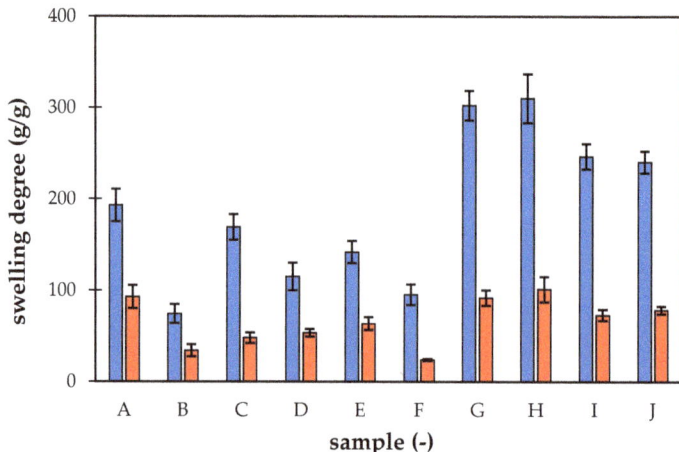

Figure 7. Swelling degree of prepared superabsorbent polymers in deionized (blue) and tap (red) water. Samples A and B contain only NPK, C, D, E, and F combination of NPK and LH, G and H only LH. Samples I and J are without nutrients. AM was added to samples E, F, G, and J.

It was confirmed that the presence of acrylamide (samples E, F, and G) had little impact on the swelling ability of the studied hydrogels both in deionized and tap water. The samples without the presence of AM exhibited no lowering of the water absorbency. When we compare "pure" hydrogels differing only in the AM addition without NPK and LH (samples I and J), no difference in their swelling behaviors was observed. Some authors [53,62–64] studied the effect of AM content on the water absorbency and observed a maximum for different AM/AA ratios (0.1 [64], 0.4 [62], and 1 [53]). In this work, the AM addition is lower in comparison with published works [2,3,5,10] because of its toxicity [41–44]. The aim was to investigate potential positive effects on the prepared superabsorbent polymers, which were not confirmed in the case of water absorbency.

2.4. Rheological Properties of Prepared Superabsorbent Polymers

In Figure 8, the dynamic moduli of all superabsorbent polymers enriched by a fertilizer are compared. We can see that the storage modulus G' proportional to the elastic component of the hydrogel is higher than loss modulus G'' proportional to the viscous component of the hydrogel. It means that all prepared materials have an elastic character. Storage modulus G' increases with increasing frequency (samples A, C, E, F, G, and H) or seems to be practically independent of frequency (samples B and D). Samples B and D contain higher amounts of NPK, and no acrylamide was used in their preparation. When AM was added (sample F), the frequency dependence of G'' becomes increasing. A general effect of a higher amount of NPK is, therefore, the increase in (elastic) storage modulus (samples B, D, F). The increase in the difference between G' and G'' for samples without AM and the decrease in (viscous) loss modulus G'' with increasing frequency supported the elastic character of prepared hydrogels (Figure 8a,b). Superabsorbent polymers containing lignohumate had lower dynamic moduli G' and G'' (Figure 8b–d) in comparison with samples A and B enriched only by NPK (Figure 8a). The character of frequency dependence of loss modulus G'' changed with the LH addition. In contrast to Figure 8a, loss modulus increased slightly with increasing frequency or remained practically constant (except sample D with higher NPK content which kept decreasing G'' dependence). The lowest values of both moduli had samples G and H containing only lignohumate without NPK. The addition of acrylamide into superabsorbent polymers resulted in a stronger increase of storage modulus G' with increasing frequency (compare samples D and F in Figure 8b,c) or in the decrease of both moduli in Figure 8d.

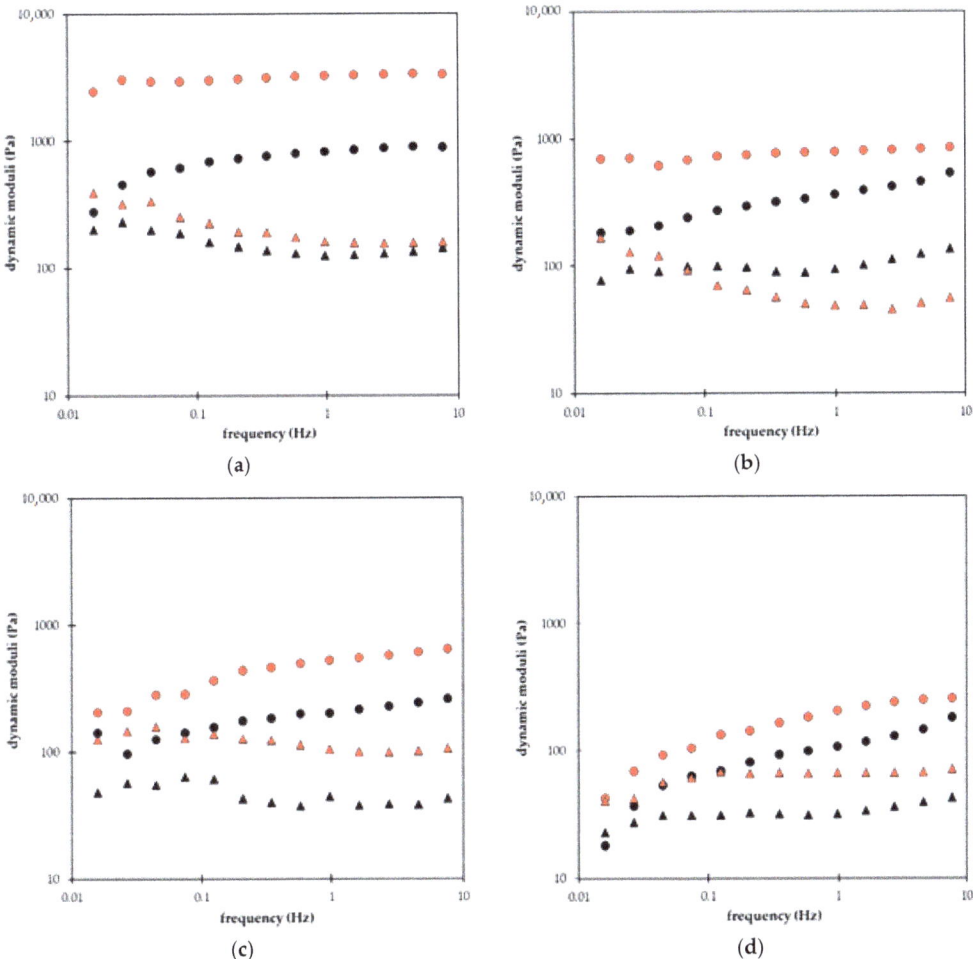

Figure 8. Storage modulus G' (circles) and loss modulus G'' (triangles) in the dependence of frequency. Comparison of different NPK contents without LH (**a**): samples A (black) and B (red). Comparison of different NPK contents with LH (**b**): samples C (black) and D (red). Effect of AM addition to polymers containing NPK and LH (**c**): samples E (black) and F (red). Effect of AM addition to polymers containing LH without NPK (**d**): samples G (black) and H (red).

In summary, we can state several conclusions about how the individual constituents affect the viscoelastic behavior of superabsorbent polymers: (1) the increase in NPK content resulted in the increase in (elastic) storage modulus G' and decreasing character of frequency dependence of (viscous) loss modulus G''; (2) the addition of LH caused the decrease in both moduli G' and G''; (3) the addition AM resulted in the decrease in both moduli G' and G''. It means that the NPK fertilizer supported the elastic character of prepared materials while the addition of lignohumate shifted the rheological behavior of superabsorbent polymers to more liquid. The more elastic character of samples with higher content of NPK means that they can resist mechanical stress, and their structure can be worse damaged. On the other hand, this effect is connected with the suppression of the release of lignohumate from superabsorbent D [13]. The addition of acrylamide did not improve the properties of prepared materials. The values of both moduli decreased with AM addition. The difference between G' and G'' strongly decreased for samples with higher NPK content which resulted in a more liquid character of sample F (in comparison with sample D). The difference between both moduli for samples containing lignohumate (without NPK) remained practically the same (Figure 8d).

2.5. Stability of Prepared Superabsorbent Polymers

The stability of superabsorbent polymers was studied from three points of view. The first one was the time stability when the samples were repeatedly measured for one year. Other important aspects of the superabsorbents functionality in soil are drying/re-swelling cycles as well as the effect of freezing and defrosting of materials in winter.

An example of the time development of dynamic moduli measured for sample A is shown in Figure 9. We can see that both storage and loss moduli decreased gradually in time as materials lost their mechanical resistance. On the other hand, the decrease in both moduli was relatively favorable to the effect that the values did not differ in magnitude, and storage modulus remained above 100 Pa also after 1 year (Figure 9a). The ratio between loss and storage moduli increased slightly in time (Figure 9b). The G''/G' ratio represents a criterion of the "liquid behavior" of the studied samples. When the ratio is equal to one, the storage and loss moduli have the same values, and the degrees of liquid and elastic extents are precisely balanced. Values higher than one denote that the sample behaves as a viscoelastic liquid.

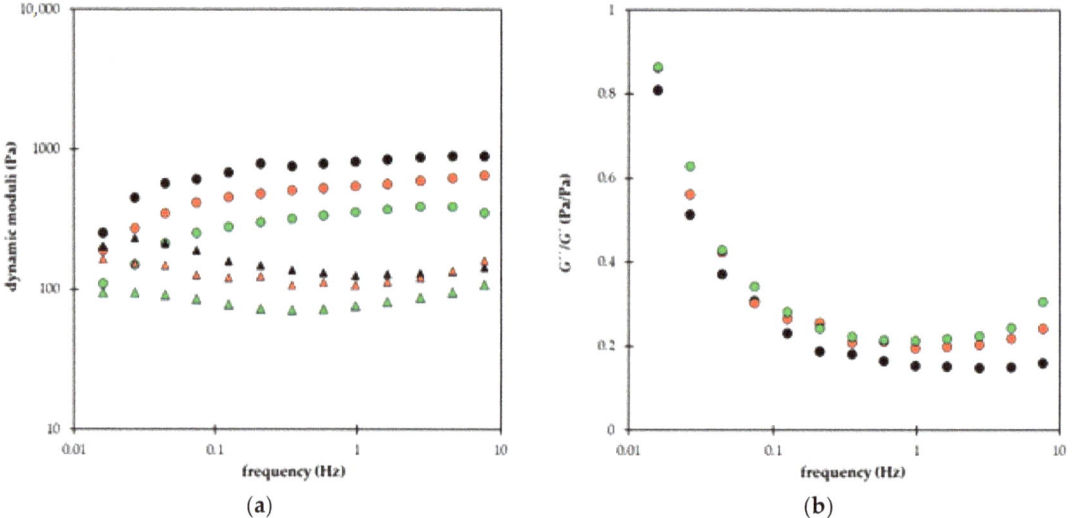

Figure 9. Time stability of sample A: Storage modulus G' (circles) and loss modulus G'' (triangles) measured for sample A (**a**) and the ratio G''/G' (**b**): after 24 h (black), 5 weeks (red), and 1 year (green).

The time stability of superabsorbent materials enriched by nutrients (NPK and LH) is compared in Figure 10. The ratio between loss and storage moduli after 24 h and 1 year are compared. The trend of a slight increase in the G''/G' ratio observed for sample A (Figure 9b) was not confirmed for all prepared materials. In some cases (samples D, E, F, G, and partially H), the ratio decreased after a year, which means that the participation of viscous character decreased. It is not caused by a strengthening of hydrogel structure but by the different rates of decrease in individual moduli, as shown in Figure 11. As can be seen, the highest decrease in both moduli was observed for samples containing NPK in combination with LH. The decrease for these samples (C, D, E, and F) was between 80 and 90%. In contrast, the decrease in moduli obtained for samples enriched with only one type of fertilizer (NPK or LH) was lower and comparable (samples A, B, G, and H). The effect of AM addition on the decrease in moduli was negligible.

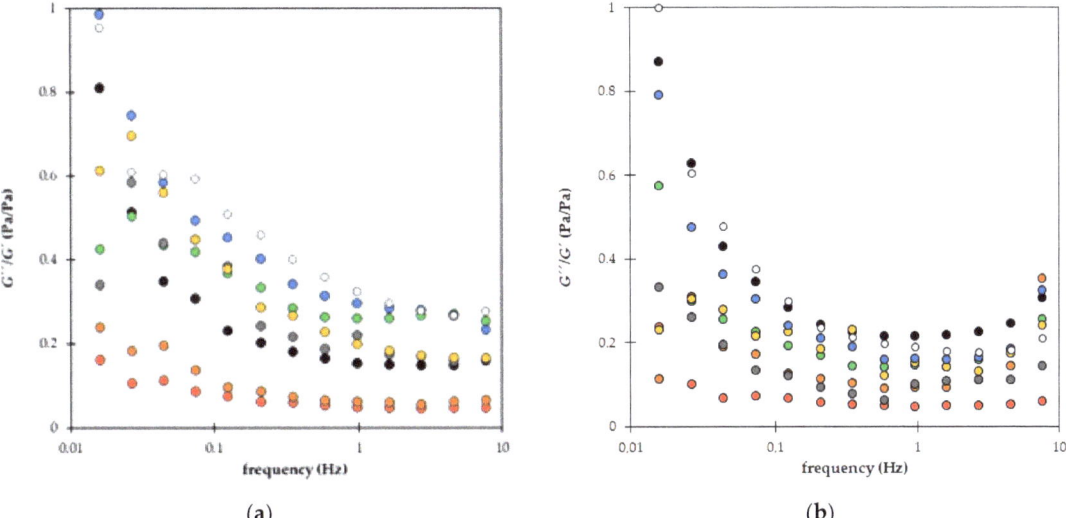

Figure 10. The ratio between loss and storage modulus after 24 h (**a**) and 1 year (**b**): samples A (black), B (red), C (green), D (orange), E (grey), F (yellow), G (blue), and H (white).

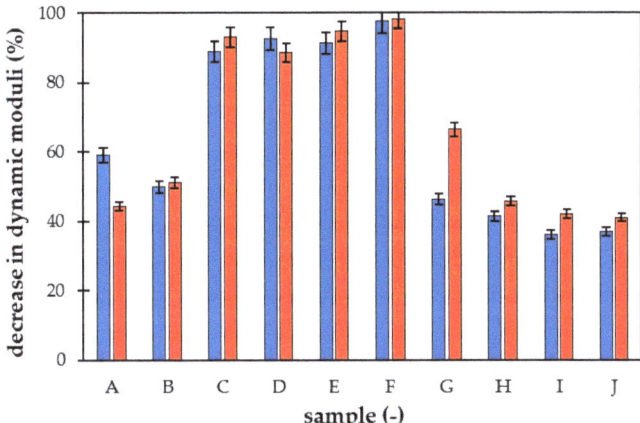

Figure 11. The average decrease in storage modulus (blue) and loss modulus (red) after 1 year. Samples A and B contain only NPK, C, D, E, and F combination of NPK and LH, G and H only LH. Samples I and J are without nutrients. AM was added to samples E, F, G, and J.

Changes in the rheological behavior of superabsorbent polymers during repeated swelling are shown in Figure 12. We can see that the G''/G' ratio decreased with the number of cycles (Figure 12a). In Figure 12a, the examples of data obtained for sample C are shown. A similar effect of re-swelling on the rheological behavior was observed for all prepared hydrogels. The viscous extent decreased gradually with a number of cycles. The results obtained for individual samples differed in the final G''/G' ratio (Figure 12b). The highest extent of liquid character was determined for samples G and H containing only lignohumate without NPK. Samples containing NPK evinced less viscous participation and seemed to be more resistant to mechanical stresses. The gradual decrease during re-swelling cycles is caused by the increase in both dynamic moduli, which was observed for all prepared hydrogels, e.g., the storage modulus of samples E and F increased after the fifth cycle more than three times. The reinforcement of prepared materials during re-swelling cycles is connected with the release of nutrients from their structure described in some refs. [13,45–47,51,65].

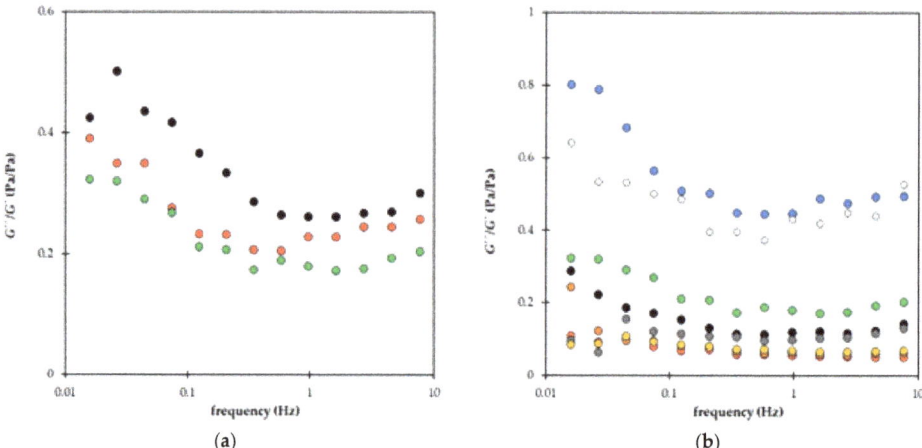

Figure 12. The ratio between loss and storage modulus after one (black), three (red), and five (green) re-swelling cycles for sample C (**a**). The G''/G' ratio after five cycles (**b**): samples A (black), B (red), C (green), D (orange), E (grey), F (yellow), G (blue), and H (white).

The effects of freezing on the rheological character of superabsorbent polymers are summarized in Figure 13. In general, the freezing and defrosting resulted in a decrease in the G''/G' ratio (except samples B and C). It means that the portion of liquid character decreased, and samples became more resistant to mechanical stresses. In contrast, the behavior of sample B shifted to a more liquid character. In this case, the reinforcement by a higher content of NPK disappeared during freezing. The ratio G''/G' remained approximately constant for samples C and D, combining the additions of NPK and lignohumate, but its value decreased if AM was added in the superabsorbent structure (samples E and F). The biggest decrease in the G''/G' ratio was observed for samples G and H containing LH without NPK. The addition of AM had a negligible effect on pure superabsorbents (samples I and J).

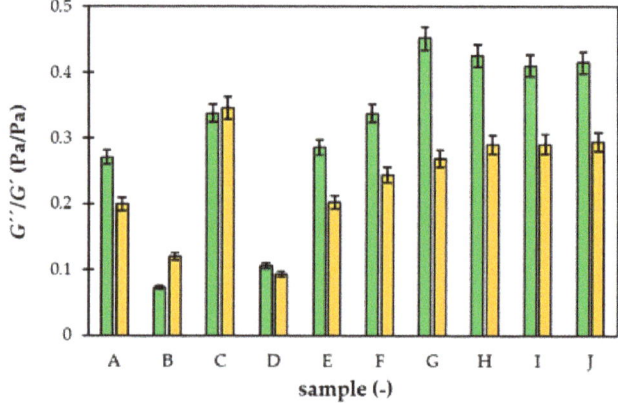

Figure 13. The average G''/G' ratio before (green) and after (yellow) freezing. Samples A and B contain only NPK, C, D, E, and F combination of NPK and LH, G and H only LH. Samples I and J are without nutrients. AM was added to samples E, F, G, and J.

3. Conclusions

Ten superabsorbent polymers based on polyacrylic acid were prepared and investigated from the point of view of their potential use in soil as a water reservoir and source of mineral and organic nutrients. The release of nutrients and the effect on plant growth were studied in detail in our previous work [13]. The aim of this work was to consider the behavior and stability of prepared materials, including repeated swelling/drying cycles and the freezing/defrosting process. The addition of NPK had a negative effect on the ability of polymers to absorb water, while the lignohumate supported the superabsorbent swelling. SEM analysis showed that the lignohumate supported the formation of complex structures with different pores and voids. The time stability was studied in one year. Both dynamic moduli decreased in time, but their ratio often remained practically the same. The highest decrease was observed for materials combining the NPK and lignohumate addition. Repeated swelling/drying cycles resulted in a decrease in the G''/G' ratio. It means that the character of materials that shifted to a more elastic and liquid character gradually decreased, which was connected with the gradual release of fertilizers. The more liquid character remained only for samples containing lignohumate without NPK. In general, the freezing and defrosting of prepared materials caused a decrease in their liquid character. Samples combined with a higher amount of NPK became more liquid, and the character of samples combining NPK with lignohumate (without acrylamide) remained the same. This was the one case where the addition of acrylamide resulted in the change of sample properties (compare samples C and D with E and F). Its effect on the pure superabsorbent polymers (without fertilizers) was negligible, and a positive effect on the utility properties of prepared materials was not confirmed.

4. Materials and Methods

4.1. Chemicals

Ten different samples of superabsorbent polymers were synthesized through the rapid solution polymerization of partially neutralized acrylic acid under normal atmospheric conditions. Some samples were based on acrylic acid and acrylamide mixture. Powders of potassium lignohumate (Amagro s. r. o., Prague, Czech Republic) and NPK 20-8-8 (Lovochemie a.s., Lovosice, Czech Republic) were used as organic and mineral nutrients.

Acrylic acid, acrylamide, N,N'-methylenebisacrylamide, and potassium peroxydisulfate were purchased from Sigma-Aldrich (St. Luis, MO, USA). KOH was purchased from Penta (Prague, Czech Republic). NPK 20-8-8 was purchased from Lovochemie a.s. (Lovosice, Czech Republic). Lignohumate was kindly provided by Amagro (Prague, Czech Republic). Its main characteristics, such as elemental composition and structural features, can be found in refs. [26,28,34,40].

4.2. Preparation of Superabsorbent Polymers

The basic materials were prepared using the following method. The weight quantity of AA (57 g) was dissolved in distilled water (100 cm^3). The AA solution (25 cm^3) was neutralized by 10 cm^3 of 8.5 M potassium hydroxide solution and crosslinked by MBA (0.016 g). Then, the initiator KPS was added (0.5 g). The mixture was continuously heated and stirred until reaching the temperature of approximately 85 °C; then, the highly viscous mixture was removed from the beaker and placed in an oven for 24 h, which was settled at 80 °C. The dried product was crushed by a hammer into small pieces [9,13]. When samples with NPK (A, B), LH (E, F), and its combination (C, D) were prepared), their powder was added to the AA solution neutralized by KOH. Alternatively, AM (0.75 g in 1.5 cm^3 of distilled water) was mixed with AA before neutralization (E, F, G, and J). The compositions of prepared samples are described in detail in Table 1.

4.3. Characterization of Superabsorbent Polymers

Prepared superabsorbent polymers were characterized by several methods: FT-IR spectrometry, scanning electron microscopy (SEM), and water absorbency.

FT-IR spectra of lignohumate, NPK, and prepared superabsorbent polymers were measured by ATR technique over the range of 4000–400 cm^{-1}. FT-IR spectrometer (Nicolet iS 50) operating with a peak resolution of 4 cm^{-1} and 128 scans were performed on each acquisition. All prepared samples were measured in a dry state before swelling.

Scanning electron microscopy (JOEL JSM-7600F, Thermo Fisher Scientific, Waltham, MA, USA) was used for the characterization of the surface morphology of prepared superabsorbent polymers before swelling and then for the characterization of their pore structure in swollen forms after freeze-drying.

Water absorbency was determined by means of swelling experiments in deionized and tap water. Superabsorbent polymers in the form of xerogel were mixed with water in the ratio of 50 mg: 100 cm^3. The swelling degree was determined on the basis of weight increase after 24 h [13]. The weight increase caused by the absorption of water was expressed as the so-called swelling degree Q. It was calculated as the amount of absorbed water (expressed as the difference between the weight of hydrogel m_h and the weight of xerogel m_x) normalized on the weight of xerogel [3–5,8,10,12,13,54]:

$$Q = \frac{m_h - m_x}{m_x}. \tag{1}$$

4.4. Rheological Properties of Prepared Superabsorbent Polymers

Rheological properties were determined by means of an Anton Paar Physica MCR 501 rheometer (Anton Paar, Gray, Austria), partially according to a previously reported method [55,56]. The measurement was performed using a parallel plate system (PP25-SN6375, 25 mm diameter) with a 1 mm gap. First, a conditioning step was performed (3 min), followed by a strain sweep test (strain: 0.01–100%, 10 rad/s) to determine the linear viscoelastic region. The region ranged from 0.01 to 1% for almost all samples, except E and G. These latter two were more pliable; their linear viscoelastic region ranged from 0.01 to 0.5%. Therefore, an amplitude of deformation of 0.1% was chosen as suitable for all further experiments (frequency sweeps).

Frequency sweep measurements were taken for all samples under the following conditions: strain, 0.1%, and frequency, 0.1–628 rad/s. Conditioning steps were performed before each measurement. Viscoelastic measurements in oscillatory shear flow—frequency sweep and strain sweep—were performed for each sample, and the obtained values of moduli G' and G'' were compared. The storage modulus G' is proportional to the extent of the elastic component, and the loss modulus G'' is proportional to the extent of the viscous component of the system. The strength of prepared samples can be characterized in summary by means of their ratio G''/G'. Three repetitions of viscoelastic measurements were performed for each sample, and the obtained values of moduli were checked for reproducibility. Three repetitions of viscoelastic measurements were performed for each sample, and the obtained values of moduli were checked for reproducibility.

4.5. Stability of Prepared Superabsorbent Polymers

The first stability study was based on the repeat measurement of rheological properties over time (up to one year). Samples were kept in a swollen state; they were stored in a desiccator with water to prevent drying out.

The next experiment was focused on the re-swelling process on superabsorbent polymers. It means that superabsorbent polymers were repeatedly dried and re-swollen in order to investigate potential changes in their properties in the cycles. The cycles were realized with deionized water as described above (Section 4.3).

Since the superabsorbent polymers should be used outside in soil, other characteristics were focused on their freezing and defrosting. Samples were frozen in their swollen form (after 1st cycle) at −19 °C (24 h) and then defrosted at laboratory temperature (24 h). Viscoelastic moduli (G' and G'') were measured after each cycle of re-swelling as well as after the defrosting of samples.

Author Contributions: Conceptualization, M.K. (Martina Klučáková) and R.K.; methodology, M.S., J.S., P.Z. and M.K. (Milan Kráčalík); validation, P.S., E.B. and M.K. (Martina Klučáková); formal analysis, M.P., E.B. and M.K. (Milan Kráčalík); writing—original draft preparation, M.K. (Martina Klučáková); project administration, M.K. (Milan Kráčalík), J.S. and M.P. All authors have read and agreed to the published version of the manuscript.

Funding: This work was supported by the project "Materials Research Centre at FCH BUT—Sustainability and Development" REG. LO1211, with financial support from National Programme for Sustainability I (Ministry of Education, Youth and Sports) and by the project AKTION Austria-Czech Republic No. 76p5 and 79p6.

Institutional Review Board Statement: Not applicable.

Informed Consent Statement: Not applicable.

Data Availability Statement: On request.

Conflicts of Interest: The authors declare no conflict of interest.

References

1. Kiatkamjornwong, S. Superabsorbent polymers and superabsorbent polymer composites. *Sci. Asia* **2007**, *33*, 39–40. [CrossRef]
2. Raju, M.P.; Raju, K.M. Design and synthesis of superabsorbent polymers. *J. Appl. Polym. Sci.* **2001**, *80*, 2635–2639. [CrossRef]
3. Raju, M.P.; Raju, K.M.; Mohan, Y.M. Synthesis and water absorbency of crosslinked superabsorbent polymers. *J. Appl. Polym. Sci.* **2002**, *85*, 1795–1801. [CrossRef]
4. Li, A.; Wang, A.; Chen, J. Studies on poly(acrylic acid)/attapulgite superabsorbent composites. II. Swelling behaviors of superabsorbent composites in saline solutions and hydrophilic solvent–water mixtures. *J. Appl. Polym. Sci.* **2004**, *94*, 1869–1876. [CrossRef]
5. Dadhaniya, P.V.; Patel, M.P.; Patel, R.G. Swelling and dye adsorption study of novel superswelling [Acrylamide/N-vinylpyrrolidone/3(2-hydroxyethyl carbamoyl) acrylic acid] hydrogels. *Polym. Bull.* **2006**, *57*, 21–31. [CrossRef]
6. Nnadi, F.; Brave, C. Environmentally friendly superabsorbent polymers for water conservation in agricultural lands. *J. Soil Sci. Environ. Manag.* **2011**, *2*, 206–211.
7. Pó, R. Water-absorbent polymers: A patent survey. *J. Macromol. Sci. Polymer. Rev.* **1994**, *34*, 607–662. [CrossRef]
8. Li, A.; Zhang, J.; Wang, A. Synthesis, characterization and water absorbency properties of poly(acrylic acid)/sodium humate superabsorbent composite. *Polym. Adv. Technol.* **2005**, *16*, 675–680. [CrossRef]
9. Chu, M.; Zhu, S.Q.; Li, H.M.; Huang, Z.B.; Li, S.Q. Synthesis of poly(acrylic acid)/sodium humate superabsorbent composite for agricultural use. *J. Appl. Polym. Sci.* **2006**, *102*, 5137–5143. [CrossRef]
10. Chu, M.; Zhu, S.Q.; Huang, Z.B.; Li, H.M. Influence of potassium humate on the swelling properties of a poly(acrylic acid-co-acrylamide)/potassium humate superabsorbent composite. *J. Appl. Polym. Sci.* **2008**, *107*, 3727–3733. [CrossRef]
11. Liu, J.; Wang, Q.; Wang, A. Synthesis and characterization of chitosan-g-poly(acrylic acid)/sodium humate superabsorbent. *Carbohydr. Polym.* **2007**, *70*, 166–173. [CrossRef]
12. Gao, L.; Wang, S.; Zhao, X. Synthesis and characterization of agricultural controllable humic acid superabsorbent. *J. Environ. Sci.* **2013**, *25*, S69–S76. [CrossRef] [PubMed]
13. Kratochvílová, R.; Sedláček, P.; Pořízka, J.; Klučáková, M. Composite materials for controlled release of mineral nutrients and humic substances for agricultural application. *Soil Use Manag.* **2021**, *37*, 460–467. [CrossRef]
14. Zhang, D.; Tang, Y.; Zhang, C.; Huhe, F.N.U.; Wu, B.; Gong, X.; Chuang, S.S.C. Formulating zwitterionic, responsive polymers for designing smart soils. *Nano Micro Small* **2022**, *18*, 2203899. [CrossRef]
15. Pushpamalar, J.; Langford, S.J.; Ahmad, M.B.; Lim, Y.Y.; Hashim, K. Eco-friendly smart hydrogels for soil conditioning and sustain release fertilizer. *Int. J. Environ. Sci. Technol.* **2018**, *15*, 2059–2074. [CrossRef]
16. Kabir, S.M.F.; Sikdar, P.P.; Haque, B.; Bhuiyan, M.A.R.; Ali, A.; Islam, M.N. Cellulose-based hydrogel materials: Chemistry, properties and their prospective applications. *Prog. Biomater.* **2018**, *7*, 153–174. [CrossRef]
17. Sikdar, P.; Uddin, M.M.; Dip, T.M.; Islam, S.; Hoque, M.S.; Dhar, A.K.; Wu, S. Recent advances in the synthesis of smart hydrogels. *Mater. Adv.* **2021**, *2*, 4532–4573. [CrossRef]
18. Azeem, M.K.; Islam, A.; Khan, R.U.; Rasool, A.; Qureshi, M.A.R.; Rizwan, M.; Sher, F.; Rasheed, T. Eco-friendly three-dimensional hydrogels for sustainable agricultural applications: Current and future scenarios. *Polym. Adv. Technol.* **2023**. (first published). [CrossRef]
19. Chakraborty, R.; Mukhopadhyay, A.; Paul, S.; Sarkar, S.; Mukhopadhyay, R. Nanocomposite-based smart fertilizers: A boon to agricultural and environmental sustainability. *Sci. Total Environ.* **2023**, *863*, 160859. [CrossRef] [PubMed]
20. Ramazani-Harandi, M.J.; Zohuriaan-Mehr, M.J.; Yousefi, A.A.; Ershad-Langroudi, A.; Kabiri, K. Effects of structural variables on AUL and rheological behavior of SAP gels. *J. Appl. Polym. Sci.* **2009**, *113*, 3676–3686. [CrossRef]

21. Zaharia, A.; Radu, A.L.; Iancu, S.; Florea, A.M.; Sandu, T.; Minca, I.; Fruth-Oprisan, V.; Teodorescu, M.; Sarbu, A.; Iordache, T.V. Bacterial cellulose-poly(acrylic acid-co-N,N'-methylene-bis-acrylamide) interpenetrated networks for the controlled release of fertilizers. *RSC Adv.* **2018**, *8*, 17635. [CrossRef] [PubMed]
22. Rehman, T.U.; Shah, L.A. Rheological investigation of polymer hydrogels for industrial application: A review. *Int. J. Polym. Anal. Charact.* **2022**, *27*, 430–445. [CrossRef]
23. Zhang, Q.; Zhou, W.; Liang, G.; Wang, X.; Sun, J.; He, P.; Li, L. Effects of different organic manures on the biochemical and microbial characteristics of albic paddy soil in a short-term experiment. *PLoS ONE* **2015**, *10*, 0124096. [CrossRef] [PubMed]
24. Vishwanath, V.; Kumar, S.; Purakayastha, T.J.; Datta, S.P.; Rosin, K.G.; Mahapatra, P.; Sinha, S.K.; Yadav, S.P. Impact of forty-seven years of long-term fertilization and liming on soil health, yield of soybean and wheat in an acidic Alfisol. *Arch. Agron. Soil Sci.* **2022**, *68*, 531–546. [CrossRef]
25. Mi, W.H.; Sun, Y.; Xia, S.; Zhao, H.T.; Mi, W.T.; Brookes, P.C.; Liu, Y.; Wu, L. Effect of inorganic fertilizers with organic amendments on soil chemical properties and rice yield in a low-productivity paddy soil. *Geoderma* **2018**, *320*, 23–29. [CrossRef]
26. Novák, F.; Šestauberová, M.; Hrabal, R. Structural features of lignohumic acids. *J. Mol. Struct.* **2015**, *1093*, 179–185. [CrossRef]
27. Holub, P.; Klema, K.; Tuma, I.; Vavríková, J.; Surá, K.; Veselá, B.; Urban, O.; Záhora, J. Application of organic carbon affects mineral nitrogen uptake by winter wheat and leaching in subsoil: Proximal sensing as a tool for agronomic practice. *Sci. Total Environ.* **2020**, *717*, 137058. [CrossRef]
28. Klučáková, M.; Kalina, M.; Enev, V. How the supramolecular nature of lignohumate affects its diffusion in agarose hydrogel. *Molecules* **2020**, *25*, 5831. [CrossRef]
29. Klučáková, M. Complexation of metal ions with solid humic acids, humic colloidal solutions, and humic hydrogel. *Environ. Eng. Sci.* **2014**, *31*, 612–620. [CrossRef]
30. Klučáková, M.; Kalina, M. Composition, particle size, charge and colloidal stability of pH-fractionated humic acids. *J. Soil. Sediment.* **2015**, *15*, 1900–1908. [CrossRef]
31. Klučáková, M. Dissociation properties and behavior of active humic fractions dissolved in aqueous systems. *React. Funct. Polym.* **2016**, *109*, 9–14. [CrossRef]
32. Klučáková, M. Characterization of pH-fractionated humic acids with respect to their dissociation behaviour. *Environ. Sci. Pollut. Res.* **2016**, *23*, 7722–7731. [CrossRef] [PubMed]
33. Klučáková, M. Conductometric study of the dissociation behavior of humic and fulvic acids. *React. Funct. Polym.* **2018**, *128*, 24–28. [CrossRef]
34. Enev, V.; Pospíšilová, L.; Klučáková, M.; Liptaj, T.; Doskočil, L. Spectral characterization of selected natural humic substances. *Soil Water Res.* **2014**, *9*, 9–17. [CrossRef]
35. Vuorinen, I.; Hamberg, L.; Müller, M.; Seiskari, P.; Pennanen, T. Development of growth media for solid substrate propagation of ectomycorrhiza fungi for inoculation of Norway spruce (*Picea abies*) seedlings. *Mycorrhiza* **2015**, *25*, 311–324. [CrossRef] [PubMed]
36. Adani, F.; Genevini, P.; Zaccheo, P.; Zocchi, G. The effect of commercial humic acid on tomato plant growth and mineral nutrition. *J. Plant Nutr.* **1998**, *21*, 561–575. [CrossRef]
37. Arancon, N.Q.; Edwards, C.A.; Bierman, P.; Welch, C.; Metzger, J.D. Influences of vermicomposts on field strawberries: 1. Effects on growth and yields. *Bioresour. Technol.* **2004**, *93*, 145–153. [CrossRef]
38. Arancon, N.Q.; Edwards, C.A.; Bierman, P.; Metzger, J.D.; Lucht, C. Effects of vermicomposts produced from cattle manure, food waste and paper waste on the growth and yield of peppers in the field. *Pedobiologia* **2005**, *49*, 297–306. [CrossRef]
39. Arancon, N.Q.; Edwards, C.A.; Bierman, P. Influences of vermicomposts on field strawberries: 2. Effects on soil microbiological and chemical properties. *Bioresour. Technol.* **2006**, *97*, 831–840. [CrossRef]
40. Smilkova, M.; Smilek, J.; Kalina, M.; Sedláček, P.; Pekař, M.; Klučáková, M. A simple technique for assessing of the cuticular diffusion of humic acid biostimulants. *Plant Methods* **2019**, *15*, 83. [CrossRef]
41. Spencer, P.; Schaumburg, H. A review of acrylamide neurotoxicity part I. Properties, uses, and human exposure. *Can. J. Neurol. Sci.* **1974**, *1*, 143–150. [CrossRef]
42. Spencer, P.; Schaumburg, H. A review of acrylamide neurotoxicity part II. Experimental animal neurotoxicity and pathologic mechanisms. *Can. J. Neurol. Sci.* **1974**, *1*, 152–169. [CrossRef]
43. Spencer, H.; Wahome, J.; Haasch, M. Toxicity evaluation of acrylamide on the early life stages of the zebrafish embryos (*Danio rerio*). *J. Environ. Prot.* **2018**, *9*, 1082–1091. [CrossRef]
44. Matoso, V.; Bargi-Souza, P.; Ivanski, F.; Romano, M.A.; Romano, R.M. Acrylamide: A review about its toxic effects in the light of Developmental Origin of Health and Disease (DOHaD) concept. *Food Chem.* **2019**, *283*, 422–430. [CrossRef]
45. Rashidzadeh, A.; Olad, A. Slow-released NPK fertilizer encapsulated by NaAlg-g-poly(AA-co-AAm)/MMT superabsorbent nanocomposite. *Carbohydr. Polym.* **2014**, *114*, 269–278. [CrossRef] [PubMed]
46. Olad, A.; Zebhi, H.; Salari, D.; Mirmohseni, A.; Tabar, A.R. Slow-release NPK fertilizer encapsulated by carboxymethyl cellulose-based nanocomposite with the function of water retention in soil. *Mater. Sci. Eng. C* **2018**, *90*, 333–340. [CrossRef] [PubMed]
47. Lee, Y.N.; Ahmed, O.H.; Wahid, S.A.; Jalloh, M.B.; Muzah, A.A. Nutrient release and ammonia volatilization from biochar-blended fertilizer with and without densification. *Agronomy* **2021**, *11*, 2082. [CrossRef]
48. Senesi, N.; D'Orazio, V.; Ricca, G. Humic acids in the first generation of Eurosoils. *Geoderma* **2003**, *116*, 325–344. [CrossRef]
49. Feng, L.; Yang, H.; Dong, X.; Lei, H.; Chen, D. pH-sensitive polymeric particles as smart carriers for rebar inhibitors delivery in alkaline condition. *J. Appl. Polym. Sci.* **2018**, *135*, 45886. [CrossRef]

50. Zhang, W.; Liu, Q.; Xu, Y.; Mu, X.; Zhang, H.; Lei, Z. Waste cabbage-integrated nutritional superabsorbent polymers for water retention and absorption applications. *Langmuir* **2022**, *38*, 14869–14878. [CrossRef]
51. Ramli, R.A.; Lian, Y.M.; Nor, N.M.; Azman, N.I.Z. Synthesis, characterization, and morphology study of coco peat-grafted-poly(acrylic acid)/NPK slow release fertilizer hydrogel. *J. Polym. Res.* **2019**, *26*, 266. [CrossRef]
52. Singh, T.; Singhal, R. Poly(acrylic acid/acrylamide/sodium humate) superabsorbent hydrogels for metal ion/dye adsorption: Effect of sodium humate concentration. *J. Appl. Polym. Sci.* **2012**, *125*, 1267–1283. [CrossRef]
53. Tomar, R.S.; Gupta, I.; Singhal, R.; Nagpal, A.K. Synthesis of poly (acrylamide-co-acrylic acid) based superabsorbent hydrogels: Study of network parameters and swelling behaviour. *Polym. Plast. Technol. Eng.* **2007**, *46*, 481–488. [CrossRef]
54. Wu, J.; Lin, J.; Zhou, M.; Wei, C. Synthesis and properties of starch-graft-polyacrylamide/clay superabsorbent composite. *Macromol. Rapid Commun.* **2000**, *21*, 1032–1034. [CrossRef]
55. Seetapan, N.; Wongsawaeng, J.; Kiatkamjornwong, S. Gel strength and swelling of acrylamide-protic acid superabsorbent copolymers. *Polym. Advan. Technol.* **2011**, *22*, 1685–1695. [CrossRef]
56. Pourjavadi, A.; Hosseinzadeh, H. Synthesis and properties of partially hydrolyzed acrylonitrile-co-acrylamide superabsorbent hydrogel. *Bull. Korean Chem. Soc.* **2010**, *31*, 3136–3172. [CrossRef]
57. Kabiri, K.; Omidian, H.; Hashemi, S.A.; Zohuriaan-Mehr, M.J. Synthesis of fast-swelling superabsorbent hydrogels: Effect of crosslinker type and concentration on porosity and absorption rate. *Eur. Polym. J.* **2003**, *39*, 1341–1348. [CrossRef]
58. Kabiri, K.; Zohuriaan-Mehr, M.J. Superabsorbent hydrogel composites. *Polym. Adv. Technol.* **2003**, *14*, 438–444. [CrossRef]
59. Shahid, S.; Qidwai, A.; Anwar, F.; Ullah, I.; Rashid, U. Effects of a novel poly (AA-co-AAm)/AlZnFe$_2$O$_4$/potassium humate superabsorbent hydrogel nanocomposite on water retention of sandy loam soil and wheat seedling growth. *Molecules* **2012**, *17*, 12587–12602. [CrossRef]
60. Nada, W.M.; Blumenstei, O. Characterization and impact of newly synthesized superabsorbent hydrogel nanocomposite on water retention characteristics of sandy soil and grass seedling growth. *J. Soil Sci.* **2015**, *10*, 153–165. [CrossRef]
61. Bai, W.; Zhang, H.; Liu, B.; Wu, Y.; Song, J. Effects of super-absorbent polymers on the physical and chemical properties of soil following different wetting and drying cycles. *Soil Use Manag.* **2010**, *26*, 253–260. [CrossRef]
62. Zhang, X.; Wang, X.; Li, L.; Zhang, S.; Wu, R. Preparation and swelling behaviors of a high temperature resistant superabsorbent using tetraallylammonium chloride as crosslinking agent. *React. Funct. Polym.* **2015**, *87*, 15–21. [CrossRef]
63. Pourjavadi, A.; Bardajee, G.R.; Soleyman, R. Synthesis and swelling behavior of a new superabsorbent hydrogel network based on polyacrylamide grafted onto salep. *J. Appl. Polym. Sci.* **2009**, *112*, 2625–2633. [CrossRef]
64. He, G.; Ke, W.; Chen, X.; Kong, Y.; Zheng, H.; Yin, Y.; Cai, W. Preparation and properties of quaternary ammonium chitosan-g-poly(acrylic acid-co-acrylamide) superabsorbent hydrogels. *React. Funct. Polym.* **2017**, *111*, 14–21. [CrossRef]
65. Zheng, Y.; Gao, T.; Wang, A. Preparation, swelling, and slow-release characteristics of superabsorbent composite containings humate. *Ind. Eng. Chem. Res.* **2008**, *47*, 1766–1773. [CrossRef]

Disclaimer/Publisher's Note: The statements, opinions and data contained in all publications are solely those of the individual author(s) and contributor(s) and not of MDPI and/or the editor(s). MDPI and/or the editor(s) disclaim responsibility for any injury to people or property resulting from any ideas, methods, instructions or products referred to in the content.

Article

Poly(acrylic acid-co-acrylamide)/Polyacrylamide pIPNs/Magnetite Composite Hydrogels: Synthesis and Characterization

Marin Simeonov [1,*], Anton Atanasov Apostolov [1], Milena Georgieva [2], Dimitar Tzankov [2] and Elena Vassileva [1]

[1] Laboratory on Structure and Properties of Polymers, Faculty of Chemistry and Pharmacy, University of Sofia "St. Kliment Ohridski", 1, James Bourchier blvd., 1164 Sofia, Bulgaria
[2] Faculty of Physics, University of Sofia "St. Kliment Ohridski", 5, James Bourchier blvd., 1164 Sofia, Bulgaria
* Correspondence: m.simeonov@chem.uni-sofia.bg

Citation: Simeonov, M.; Apostolov, A.A.; Georgieva, M.; Tzankov, D.; Vassileva, E. Poly(acrylic acid-co-acrylamide)/Polyacrylamide pIPNs/Magnetite Composite Hydrogels: Synthesis and Characterization. *Gels* **2023**, *9*, 365. https://doi.org/10.3390/gels9050365

Academic Editors: Dong Zhang, Jintao Yang, Xiaoxia Le and Dianwen Song

Received: 11 March 2023
Revised: 20 April 2023
Accepted: 22 April 2023
Published: 26 April 2023

Copyright: © 2023 by the authors. Licensee MDPI, Basel, Switzerland. This article is an open access article distributed under the terms and conditions of the Creative Commons Attribution (CC BY) license (https://creativecommons.org/licenses/by/4.0/).

Abstract: Novel composite hydrogels based on poly(acrylic acid-co-acrylamide)/polyacrylamide pseudo-interpenetrating polymer networks (pIPNs) and magnetite were prepared via in situ precipitation of Fe^{3+}/Fe^{2+} ions within the hydrogel structure. The magnetite formation was confirmed by X-ray diffraction, and the size of the magnetite crystallites was shown to depend on the hydrogel composition: the crystallinity of the magnetite particles increased in line with PAAM content within the composition of the pIPNs. The Fourier transform infrared spectroscopy revealed an interaction between the hydrogel matrix, via the carboxylic groups of polyacrylic acid, and Fe ions, which strongly influenced the formation of the magnetite articles. The composites' thermal properties, examined using differential scanning calorimetry (DSC), show an increase in the glass transition temperature of the obtained composites, which depends on the PAA/PAAM copolymer ratio in the pIPNs' composition. Moreover, the composite hydrogels exhibit pH and ionic strength responsiveness as well as superparamagnetic properties. The study revealed the potential of pIPNs as matrices for controlled inorganic particle deposition as a viable method for the production of polymer nanocomposites.

Keywords: magnetite; polyacrylamide; poly(acrylic acid); interpenetrating polymer networks; hydrogels; polymer nanocomposites

1. Introduction

Hydrogels are three-dimensional networks built of hydrophilic polymers which can absorb and retain large amounts of water. Due to their high water content and soft nature, they are very similar to living tissues and thus possess very good biocompatibility. Therefore, they are extensively studied for biomedical applications such as drug delivery systems [1], biosensors [2], tissue engineering [3], etc. The combination of different polymers has the potential to ensure the development of novel polymeric materials with improved functionality. One of the possible ways to combine polymers of different natures and with different properties in one material is the formation of interpenetrating polymer networks (IPNs). IPNs comprise two or more networks that are at least partially interlaced on a molecular scale but not covalently bonded to each other and cannot be separated unless chemical bonds are broken [4]. When the IPN comprises a polymer network formed in the presence of a linear polymer, the so-called pIPNs are obtained [5].

Magnetite (Fe_3O_4) particles are used for various applications such as drug delivery, and bioseparation, and as contrast agents [6]. The major problem which arises when processing magnetic particles is their colloidal instability and tendency to aggregate due to dipole–dipole interactions [7]. To overcome their aggregation, the magnetite particles are either coated with polymers [8,9] or surface functionalized, e.g., with acids [10] or bioactive compounds [11]. Another possible approach to address this challenge is the in situ formation of magnetic particles within a polymer matrix, e.g., in a hydrogel [9,12,13], the latter being used as a template which also provides the possibility of controlling the particles' size as well as the size distribution. This approach is inspired by nature as it is known that some bacteria are able to synthesize magnetic particles intracellularly using the in situ approach [14].

Poly(acrylic acid) (PAA) and polyacrylamide (PAAM) are among the polymers used for coating magnetite particles [8] due to their functionality. Both PAA and PAAM are known to interact with iron ions [9,10], providing centers for nucleation and crystal growth via their functional groups. When PAA and PAAM are combined in one material, e.g., as in copolymers or IPNs, hydrogen bonds are formed between their pendant carboxylic and amide groups, respectively. Thus, the PAA/PAAM-based materials are known to exhibit upper critical solution temperature (UCST) behavior due to these hydrogen bonds. These materials can respond simultaneously to changes in environmental pH (due to the COOH groups in the PAA) and temperature (due to the hydrogen bonds between both polymers), which defines them as smart materials.

Linear PAA was applied as a coating to stabilize magnetite nanoparticles [8,15] and prevent their agglomeration. Sanchez et al. study the effect of molecular weight and PAA content on the magnetic and structural properties of iron oxide nanoparticles. The results show that the increase in PAA content results in smaller sizes and narrower size distributions of the synthesized iron oxide particles. Magnetization analysis shows that the PAA-coated particles are superparamagnetic [8]. Chełminiak et. al. synthesized PAA-stabilized magnetite nanoparticles through photopolymerization. The synthesized particles were characterized by their pH sensitivity and superparamagnetic properties, which allow their fast separation from the media [15]. Several studies show that the PAA-coated magnetic particles can be useful materials for biomedical applications. The immobilization of PAA potentiates the anticoagulant activity and increases the thrombin time [16] as well as providing good biocompatibility [17] depending on the magnetic particles' sizes. Large particles with high PAA content cause a reduction in proliferation of cell cultures, while smaller particles are not harmful to cell proliferation. PAAM hydrogels were explored as matrices for in situ formation of magnetite [12,13]. It was demonstrated that the water uptake and drug release rate of these hydrogels can be controlled by applying an external magnetic field. This is due to the alignment of magnetic moments of individual magnetite nanoparticles, which leads to the hydrogel swelling at a faster rate [12]. A current study demonstrates the coating of magnetite particles with a PAA/PAAM copolymer using gamma radiation. The pH- and temperature-responsiveness of the resulting hydrogels appear to depend on the hydrogels' composition, i.e., the PAA/PAAM ratio. The resulting materials appear to be superparamagnetic. These hydrogels were successfully tested as drug delivery systems for doxorubicin [18].

To the best of our knowledge, however, no study to date has been dedicated to the role of IPN hydrogels combining both polymers in the in situ formation of magnetic particles. Thus, the aim of the study was to evaluate the role of P(AA-co-AAM)/PAAM pIPNs for the in situ formation of magnetite particles and therefore obtain novel P(AA-co-AAM)/PAAM pIPNs/magnetite composite hydrogels. We assume that the pIPNs' composition and structure allow the control of the overall polymer network density and functionality and hence they are expected to influence the size and size distribution of the magnetite particles deposited in situ when compared to the single PAA and PAAM networks.

2. Results and Discussion

2.1. Swelling Behavior

Equilibrium Swelling Ratio (ESR)

The neat pIPNs hydrogels increase their initial weight ~20 times upon swelling in water (ESR~20–24) but no clear dependence of their ESR on the PAA/PAAM ratio is seen (Figure 1A). The pIPNs' ESR values are comparable to the ESR of the AA100 sample and higher when compared to the neat PAAM (ESR of AA0~18). Nevertheless, such an ESR value is comparatively low for super absorbents such as PAA and PAAM. This could be because, at pH~6 (the swelling experiment is performed in distilled water), the prevailing number of the carboxylic groups of PAA are protonated, and ~40% of the carboxylic groups of PAA are in their anionic form [19]. Thus, two effects take place during water absorbance: (i) hydrogen bonds are formed either between –COOH groups of PAA [20] or between –COOH (from PAA) and –CONH$_2$ groups (from PAAM), which implies additional constraints on the pIPNs' swelling ability and explains the not very high ESR values; and (ii) the ionized carboxylic groups start to repulse each other and, the more carboxylic anions are formed, the stronger the repulsion, which results in the hydrogel expansion. This process lies behind the super absorbency of PAA. Thus, as the two effects act in opposite directions, no clear dependence of ESR on the pIPNs hydrogels' composition is observed (also confirmed by the applied ANOVA analysis, Table S1).

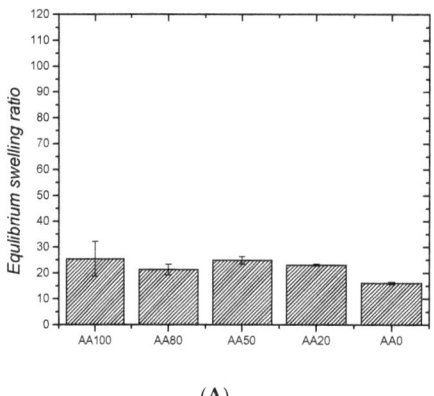
(A)

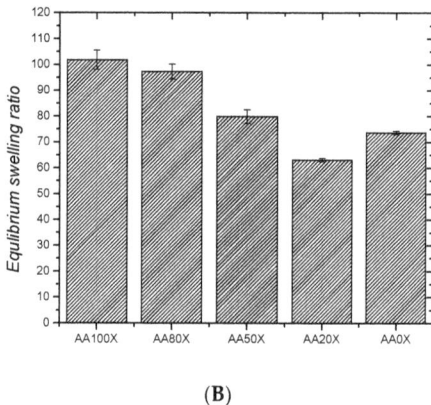
(B)

Figure 1. Equilibrium swelling ratio of: (**A**) neat pIPNs hydrogels and (**B**) pIPNs/magnetite composites.

The in situ formation of magnetite particles in the IPNs results in a significant increase in the ESR and the obtained pIPNs/magnetite composites swell much more than the neat IPNs (Figure 1B). Moreover, a clear dependence of ESR on the IPN composition is observed: the ESR decreases as the AA content decreases (Figure 1B), as also confirmed by ANOVA (Table S2 and Figures S1 and S2, Supplementary Information). This trend could be explained by the disruption of the hydrogen bonds in the neat pIPNs, due to the preferable interaction of -COOH groups from AA with Fe ions, as revealed by the ATR-IR spectra of neat IPNs (Figure S3). Thus, the secondary physical network in pIPNs is destroyed and the total network density is reduced, i.e., the swelling degree increases. In the pIPNs/magnetite composites, the carboxylic anions dictate their swelling behavior due to the repulsive interactions between adjacent -COO$^-$ anions.

We have also detected an additional ionization of the pIPNs/magnetite composites due to the PAAM alkaline hydrolysis which takes place under the experimental conditions used for magnetite formation, namely pIPNs' treatment with 6 M NaOH for 72 h (Figure S5):

$$-CONH_2 + NaOH \rightarrow -COO^- Na^+ + NH_3 \qquad (1)$$

In this way, some of the AAM monomeric units are transformed into sodium acrylate and this is confirmed when comparing the IR spectra of AA0 and AA0N (Figure S5A, Supplementary Information). Thus, the procedure for the iron oxide formation increased the number of -COO$^-$ anions (Table S3, Figure S5) resulting in an enhanced electrostatic repulsion between adjacent chains and contributing to the increased ESR of pIPNs/magnetite composites compared to that of the neat pIPNs. The ESR of pIPNs/magnetite composites increases at higher PAA content, which is expected due to the enhanced number of COO-anions and the repulsion between adjacent anions in the pIPNs, resulting in both cases in a higher swelling degree.

When comparing the ESR of pIPNs/magnetite composites obtained using different concentrations of the Fe solutions used for the in situ iron oxide formation (namely X, Y and Z series in Materials and methods section), it is clearly seen that they do not differ in their ESR, i.e., the Fe ion concentration does not influence the ESR of pIPNs (Figure S6).

2.2. Number Average Molecular Mass between Crosslinks and Mesh Size of Neat pIPNs

As can be seen in Table 1, the polymer volume fraction $v_{2,s}$ in the swollen state increases as the AA content decreases, while the number average molecular mass between crosslinks decreases. This means that the network becomes denser as the PAA content decreases, which is in line with the ESR study of the pIPNs/magnetite composites (Figure 1B). The causes, as outlined above, are the interlaced IPN structure as well as the decrease in the number of charged groups (coming from PAA) that repulse each other and expand the polymer hydrogel upon swelling.

Table 1. P(AA-co-AAM)/PAAM pIPNs hydrogels swelling characteristics in distilled water.

Sample Designation	$v_{2,s}$ Equation (5)	$\overline{M_c}$ [Da] Equation (6)	$\overline{M_{c,t}}$ [Da] Equation (7)	ξ [nm] Equation (8)
AA100	0.0415 ± 0.0005	355,676 ± 11,809	1801	1144 ± 24
AA80	0.0464 ± 0.0037	248,403 ± 58,424	1797	862 ± 125
AA50	0.0481 ± 0.0069	211,178 ± 60,791	1790	702 ± 138
AA20	0.0484 ± 0.0059	182,124 ± 62,749	1783	561 ± 122
AA0	0.0463 ± 0.0028	166,366 ± 22,494	1777	481 ± 42

The mesh size characterizes the space between macromolecular chains, and this is the space where a solute in a network could travel as well as being the space where iron oxide nanoparticles could be formed. That is why the mesh size is expected to play an important role in the in situ formation of iron oxides. Figure 2 presents the dependence of both the number average molecular mass between crosslinks as well as the mesh size of pIPNs hydrogels on the polymer volume fraction. As the polymer volume fraction, $v_{2,s}$, increases, both Mc and ξ decrease, which again proves the network density increase. This observation is in line with similar results for polymer networks reported elsewhere [21].

The theoretical number average molecular mass between crosslinks for pIPNs hydrogels, $\overline{M_{c,t}}$, was calculated (Equation (7)) and is presented in Table 1. Both Mc values show the same dependence on pIPNs' composition, but they significantly differ in their values. The theoretically predicted $\overline{M_{c,t}}$ is much lower than the corresponding experimentally determined $\overline{M_c}$ value. This can be explained by the fact that the theoretical predictions do not consider the presence of the linear PAAM chains which contribute to the overall pIPNs samples' behavior, in particular their swelling. Following the model of Andrews et al. [22], the molecular network in a rubber sample is a dual or "hybrid" network, containing two types of chains: (1) chains which are at equilibrium when the sample is at its unstretched length (in our case these are the PAAM linear chains, part of the pIPN structure); and (2) chains which are at equilibrium when the sample is at its stretched length (this is the copolymer network). The interchain entanglements between the copolymer network and the linear polymers could possibly be the source of the significant difference between the two Mc values.

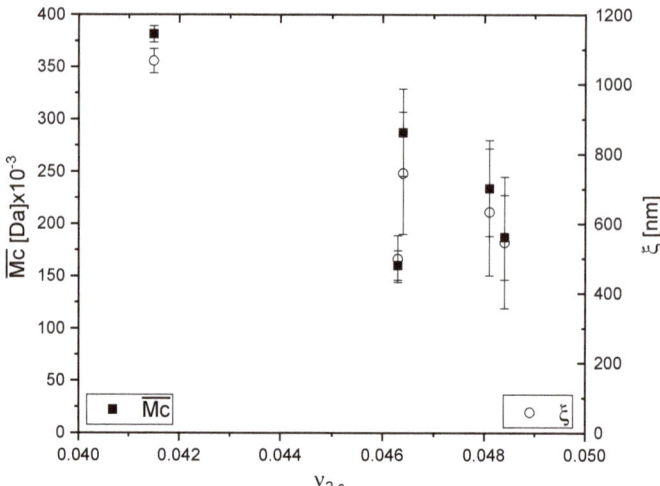

Figure 2. Dependence of the number average molecular mass between crosslinks and the mesh size of pIPNs hydrogels on polymer volume fraction.

2.3. Determination of the Iron Content in pIPNs/Magnetite Composites

The swelling ability of the pIPNs is expected to influence the quantity of iron ions that they absorb. As the amount of PAA in the pIPNs decreases, the absorbed iron increases, as revealed by two independent methods, namely flame atomic absorption spectroscopy (FAAS) and energy-dispersive X-ray analysis (EDX) (Figure 3). These results are supported by the visual appearance of the pIPNs composite hydrogels (Figure 4). The pIPNs hydrogels' color changes (deepens) from transparent through yellowish to black as PAA content in the pIPNs decreases from AA100 to AA0. Thus, the pIPNs' composition is a tool for controlling the in situ formation of magnetite within the P(AA-co-AAM)/PAAM pIPNs hydrogels. The quantity of absorbed Fe ions and hence the quantity of iron oxides formed in situ also influenced the crystallinity of the pIPNs composites. We have demonstrated that using different concentrations of the initial Fe^{3+}/Fe^{2+} solution, changing in this way the amount of in situ-formed iron oxide, changed not only the visual appearance of the samples (Figure 4c) but also the crystallinity of the formed iron oxide (Figure S9).

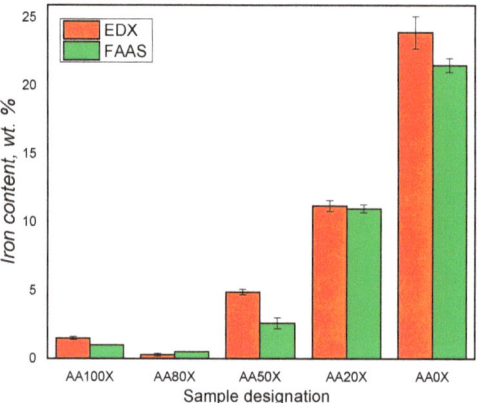

Figure 3. Iron ions content in pIPNs composites as determined by flame atomic absorption spectroscopy (FAAS) and energy-dispersive X-ray analysis (EDX).

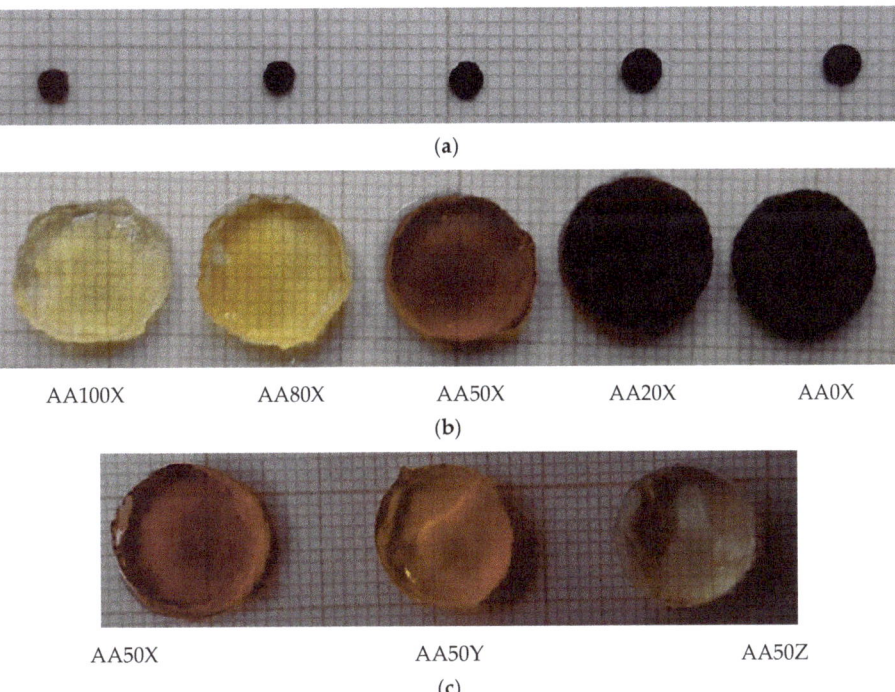

Figure 4. Appearance of the pIPNs/magnetite composites (**a**) when dry and (**b**) in their swollen state. (**c**) Comparison in the appearance of the AA50 composites obtained by swelling in Fe^{3+}/Fe^{2+} solutions with different concentrations (Z denotes the lowest Fe ions concentration used).

The higher amount of Fe in the pIPNs with the lowest PAA content could be explained by the increased amount of in situ-formed (precipitated) iron oxide particles in these pIPNs.

The precipitation of the iron oxide in AA monomeric units-rich hydrogels is hampered by the complex formation between COO^- anions from PAA and Fe cations from the salts, which is confirmed by comparing the IR spectra of AA100 (i.e., neat PAA) and one of its composites, i.e., AA100Fe (Table S4, Figure S5B). It illustrates that the bands for $-C=O$ antisym and $\nu_{C=O}$ –COOH in pIPNs observed in the spectrum of the PAA homopolymer (AA100) at 1700 cm^{-1} and 1653 cm^{-1} shift, respectively, to 1711 cm^{-1} and 1666 cm^{-1} in AA100Fe (which is the same pIPN sample swollen in Fe salts solution) due to fact that Fe ions interact preferably with –COO$^-$ groups of PAA. In contrast, the amide I, II and III bands do not change their position after swelling in Fe solution, which is clearly seen when comparing the spectra of AA0 and AA0Fe, i.e., the same network after its swelling in Fe salts aqueous solution (Table S3, Figure S5A). Thus, the interaction between Fe ions and COO$^-$ groups of PAA hampers the in situ formation of iron oxide particles. In contrast, in PAAM-rich pIPNs, the Fe ions are "free" to form iron oxide particles as the Fe ions do not interact with –CONH$_2$ groups of PAAM (Figure S5A) and thus the quantity of iron oxides formed increases.

2.4. X-ray Diffraction of the pIPNs Composites

X-ray diffraction was used to characterize the pIPNs iron oxides obtained in situ (Figure 5). The diffractograms of the pIPNs hydrogels with in situ-formed iron oxide particles show peaks which are characteristic of magnetite (at 2θ = 30.15; 35.4; 43.05; 53.5; and 57.0, marked with asterisk * in Figure 5) [23]. The hydrogels' composition, i.e., the PAA/PAAM ratio, strongly influences the magnetite formation; in the samples where AA monomeric units prevail (AA100X and AA80X), no crystal peaks are detected, while in the samples AA50X, AA20X and AA0X, the magnetite peaks are clearly seen (Figure 5). This observation is in line with the explanation provided above of how the pIPNs' composition determines the iron content in the respective pIPNs/magnetite composites. As mentioned above, the -COO$^-$ groups' interaction with Fe ions does not allow the formation of crystalline magnetite: the higher the COOH groups content, the lower the amount of crystalline magnetite.

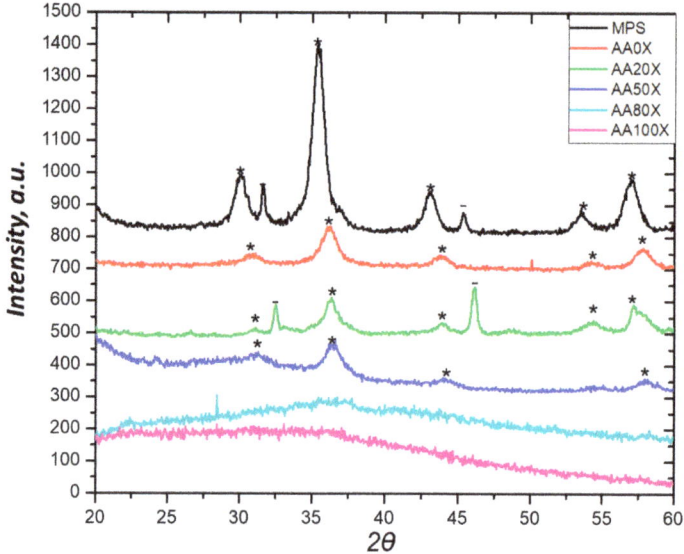

Figure 5. XRD diffractograms of pIPNs/magnetite composites.

For the sake of comparison, the X-ray diffractogram of neat iron oxide obtained using the same procedure for in situ preparation of the pIPNs/magnetite composites (designated as MPS in Table 2) is presented in Figure 5. The neat iron oxide (MPS sample) shows slightly shifted peak positions when compared to magnetite (PDF# 01-1111); these are most probably due to irregularities of the samples' surface.

Table 2. Crystallite size of magnetite particles formed in situ in pIPNs at two 2θ values.

Sample	Crystallite Size [Å] at 2θ = 35.4 D_{hkl} (311)	Crystallite Size [Å] at 2θ ~ 43 D_{hkl} (400)
AA100X	amorphous	amorphous
AA80X	11	amorphous
AA50X	46	30
AA20X	73	96
AA0X	72	90
MPS	92	98

The crystallite size of magnetite particles was calculated using the peaks at 2θ = 35.4 D_{hkl}(311) and 2θ = 43 D_{hkl}(400) (Table 4). The magnetite crystallites are very small (≤ 11 nm) which explains well the peak broadening observed in their X-ray diffractograms (Figure 5). Thus, according to the XRD results, magnetite crystallites with size ≤ 11 nm form bigger polycrystalline particles (which we call magnetite particles in the manuscript).

In Table 4, the size of MPS magnetite crystallites is also provided for the sake of comparison. As could be expected, it is higher than the crystallite size of magnetite particles formed in situ in pIPNs due to the constraints that the hydrogel exerts on the magnetite formation.

Magnetite crystallite values, such as those obtained within this study, were reported when in situ precipitation of iron oxide in a neat PAAM network was performed. Specifically, 6 and 9 nm are reported by Sivudu et al. while, without the influence of any polymer, magnetite crystallites of size ~10 nm was observed [7]. This opens up the avenue for successful exploration of pIPNs as a template for the in situ formation of magnetite as a method for the manufacture of magnetite–polymer nanocomposites. Such a method may avoid the agglomeration of magnetite nanoparticles, which usually takes place when such nanocomposites are obtained via direct blending of the polymer and the ferric oxide particles [24]. Furthermore, the size and crystallinity of magnetite particles could be controlled via pIPNs' composition as Figure 6 reveals: there is a clear relationship between the mesh size of the pIPNs and the crystallite size of the in situ-formed magnetite.

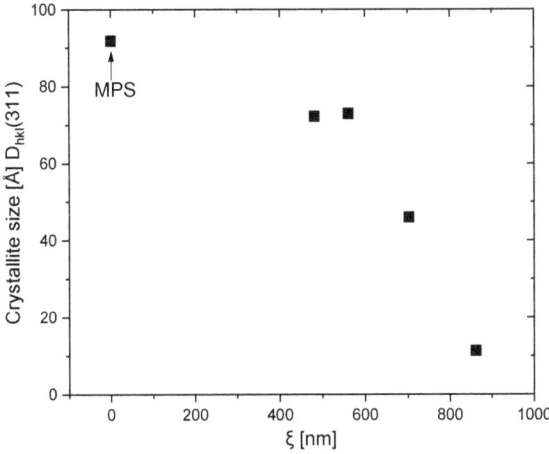

Figure 6. Dependence of the magnetite crystallite size on the pIPNs' mesh size for pIPNs/magnetite composites.

Moreover, the concentration of the initial Fe^{3+}/Fe^{2+} solution could also be used to influence the quantity of the in situ-formed iron oxides as well as their crystallinity (Figure S9). The dilution of the Fe salts solution results in an insufficient quantity of Fe ions which could be further effectively involved in the formation of crystal phases and, thus, the crystallinity could be varied.

2.5. pH-Responsive Behavior of pIPNs and the pIPNs/Magnetite Composites

The presence of –COOH groups in the pIPNs defines their ability to respond to changes in pH. At pH > pK_a^{COOH} (pK_a^{COOH} of PAA is in the range from 4.25 to 4.75) [25,26], –COOH groups are deprotonated, which results in negatively charged pIPNs. Thus, an electrostatic repulsion arises between neighboring chains and the hydrogel expands, leading to an ESR increase. This pH responsiveness is clearly seen for neat P(AA-co-AAM)/PAAM hydrogels in Figure 7A where:

- At pH = 3 (i.e., below pKa of COOH), all –COOH groups are protonated and the pIPNs hydrogels show a swelling ratio ~10 for all pIPNs compositions, i.e., AA content does not influence the ESR.
- At pH = 5 and above (i.e., above pKa of COOH), the polyanionic character of the pIPNs hydrogels makes the hydrogels increase their swelling ratio up to 70–80, i.e., 7

to 8 times, at an alkaline pH when compared to the swelling ratio obtained for pH = 3 (Figure 7A, Figure S10).

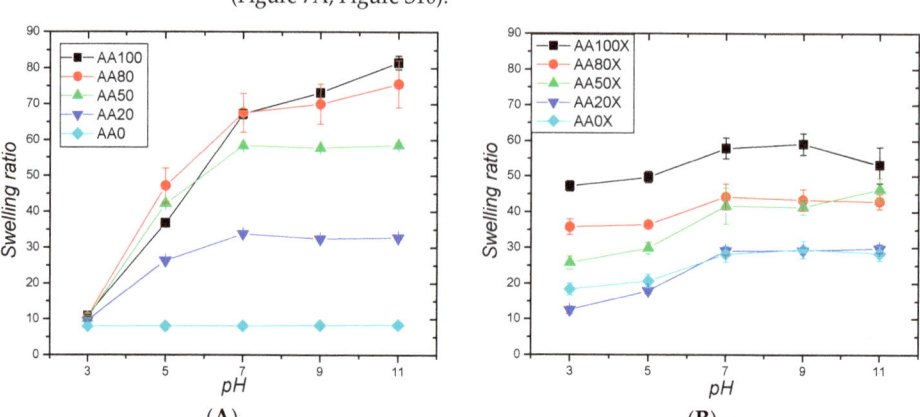

Figure 7. pH responsiveness of (**A**) neat pIPN hydrogels and (**B**) pIPNs/magnetite composite hydrogels.

The ESR value is highly dependent on the hydrogel composition: the higher the AA content, the higher the ESR. As expected, the PAAM/PAAM IPN hydrogel (AA0 sample) does not show pH responsiveness. Thus, the AA/AAM monomeric units' ratio in their copolymer is also a key tool for controlling the smart behavior of the P(AA-co-AAM)/PAAM hydrogels upon pH change.

The pIPNs/magnetite composite hydrogels also demonstrate pH responsiveness although their swelling ratio, for some compositions, increases "only" twice upon pH increase (Figure 7B). Several factors define the weaker effect of pH on the swelling ratio of pIPNs/magnetite composites. Firstly, magnetite particles act as additional crosslinking junctions, reducing the swelling ratio of the composite hydrogels in comparison to the neat pIPNs; this is clearly seen at higher pH when all COOH groups are deprotonated. Similar observations are reported for PAAM/magnetite composites where the swelling of magnetite-loaded PAAM hydrogels in physiological fluid was observed to decrease when increasing the amount of magnetite in the gel [12]. Secondly, the increased ionic strength of the media, as the pH value is adjusted using 0.1 M phosphate buffer and 0.1 M NaOH, is known to suppress the swelling of polyelectrolytes (in this particular case—PAA). Thirdly, the complexation between the carboxyl groups of PAA and iron ions partially shields the repulsion between adjacent COO^- thus diminishing the pH influence on the swelling ratio of pIPNs/magnetite composite hydrogels [27].

The different behavior of the AA0 sample (PAAM/PAAM pIPNs) and the respective composite AA0X (PAAM/PAAM pIPNs with in situ-formed magnetite) should be noted. While, as expected, AA0 shows no pH responsiveness (Figure 6A), the respective composite hydrogel exhibits some influence of pH on its swelling (Figure 6B). Similar results are reported in the literature for PAAM hydrogel loaded with magnetite nanoparticles; they also demonstrated pH responsiveness [6]. The authors explained the observed behavior by enhanced electrostatic forces between the amide groups and iron oxide. However, according to our results (Table S4, Figure S5A), which are further confirmed by a recent paper [13], this pH responsiveness could be due to the PAAM hydrolysis to PAA (Figure S4) which takes place during the magnetite formation where a high concentration of NaOH is used. Thus, COO^- anions appear along the PAAM chains which define the observed pH responsiveness of the respective composites.

2.6. Salt-Concentration Responsiveness of pIPNs and the pIPNs/Magnetite Composites

All pIPNs hydrogels—neat and composite ones—exhibit a polyelectrolyte effect, i.e., a decrease in their swelling ratio upon increasing the ionic strength of the medium (Figure 8). It is interesting to note the sharp increase in the swelling ratio (SR) of the neat pIPNs hydrogels at very low (0.001 M) NaCl concentration (Figure 8B, Table 3). The effect could be explained by the partial or complete destruction of the hydrogen bonds formed between PAA and PAAM by the stronger interaction of COOH groups with the sodium ions from NaCl. This results in liberation of part of the pIPNs chains and thus the SR sharply increases. The further increase in the ionic strength, however, shields up the repulsive interactions between carboxylic anions in the PAA and, correspondingly, the SR decreases.

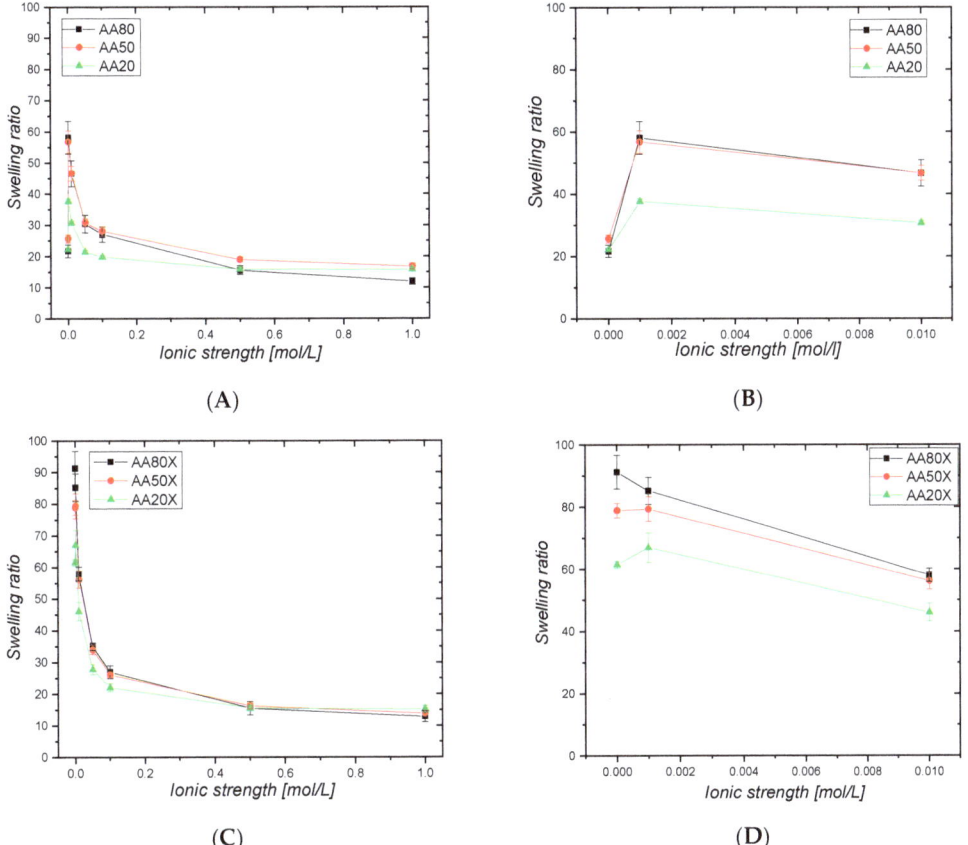

Figure 8. Ionic-strength responsiveness of pIPNs hydrogels: (**A**) and (**B**) without magnetite; (**C**) and (**D**) with in situ-formed magnetite. Figure (**B**) and (**D**) are provided in order to better illustrate the swelling ratio at ionic strength lower than 0.01 mol/L.

Table 3. Swelling ratios of pIPNs and pIPNs/magnetite composites in water and in 0.001 M NaCl aqueous solution.

Sample	H$_2$O	0.001 M NaCl
AA80	22 ± 2	58 ± 5
AA80X	91 ± 5	85 ± 4
AA50	26 ± 1	57 ± 2
AA50X	79 ± 2	79 ± 4
AA20	22 ± 1	39 ± 1
AA20X	61 ± 1	67 ± 5

For pIPNs/magnetite composite hydrogels, such SR increase at low NaCl concentrations is not observed (Figure 8C, samples with "X" in Table 3). Their SR decreases as the NaCl concentration increases most probably due to the lower number of hydrogen bonds between PAA and PAAM: the concurrent interaction of the PAA –COOH groups with Fe ions results in weaker PAA-PAAM interactions. The smaller number of hydrogen bonds in pIPNs/magnetite composites means a lower number of H-bonds that could be disrupted upon the addition of electrolytes and the resulting effect of the salt on the SR is weaker. This observation indirectly confirms the proposed explanation for the neat pIPNs' swelling behavior at low NaCl concentration (Figure 8A). Moreover, due to the PAAM hydrolysis, the number of CONH$_2$ groups decreases as they are transformed into sodium acrylates under the experimental conditions used for magnetite formation, which could also result in a decrease in the number of PAA-PAAM hydrogen bonds.

2.7. Temperature Responsiveness of pIPNs and pIPNs/Magnetite Composites

The hydrogen bonds between PAA and PAAM in the pIPNs define temperature responsiveness and PAA/PAAM-based materials show upper critical solution temperature behavior (UCST). The neat pIPNs hydrogels exhibit temperature responsiveness, which is influenced by their composition (Figure 9A): as the PAA content decreases, the PAAM content increases correspondingly, and the temperature responsiveness diminishes. The reason for this is the decreased number of inter- and intramolecular hydrogen bonds that results from variation of pIPNs composition, which defines less pronounced temperature responsiveness (Figure 9A).

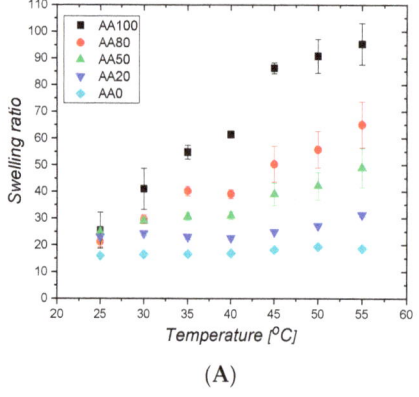

(A)

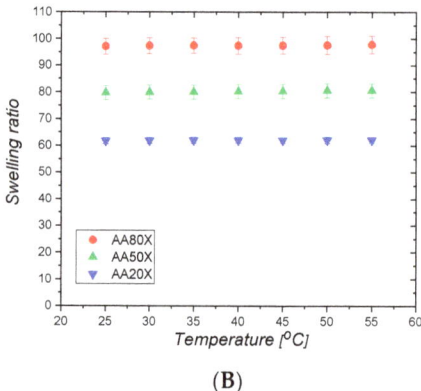

(B)

Figure 9. Temperature responsiveness of (A) neat P(AA-co-AAM)/PAAM pIPN and (B) pIPNs/magnetite hydrogels.

The pIPNs/magnetite composites, however, do not show temperature responsiveness; their swelling ratio does not change upon temperature increase (Figure 9B). This confirms the explanation above that, due to the concurrent interaction of –COOH groups from PAA with Fe ions, the number of hydrogen bonds between –COOH and –CONH$_2$ groups (from PAAM) is reduced and this results in diminishment of the temperature response. Moreover, the procedure for magnetite formation results in –CONH$_2$ groups hydrolysis, i.e., a decrease in the number of sites where hydrogen bonds could be formed, additionally reducing the number of hydrogen bonds in the composites. All these factors result in the loss of temperature responsiveness in the pIPNs/magnetite composites (Figure 9B). Similar studies with hybrid pIPNs/magnetite materials obtained at lower Fe ions concentration (Figure S11) confirm this conclusion and reveal that even low Fe concentrations could make the pIPNs lose their temperature responsiveness.

2.8. Thermal Properties

The thermal properties of neat pIPNs as well as of their composites containing magnetite were studied to evaluate the influence of magnetite on the pIPNs' properties (Figures S12 and S13). The pIPNs show one Tg which is between the Tg's of the neat PAA, namely 73 °C for the AA100 sample, and 111 °C for the neat AA0 sample (Table S6, Figure S12). The experimentally determined Tg (black squares) show a slight positive deviation from the dependence predicted by the additivity law (red dots), which could be explained by the hydrogen bonding between PAA and PAAM (Figure 10).

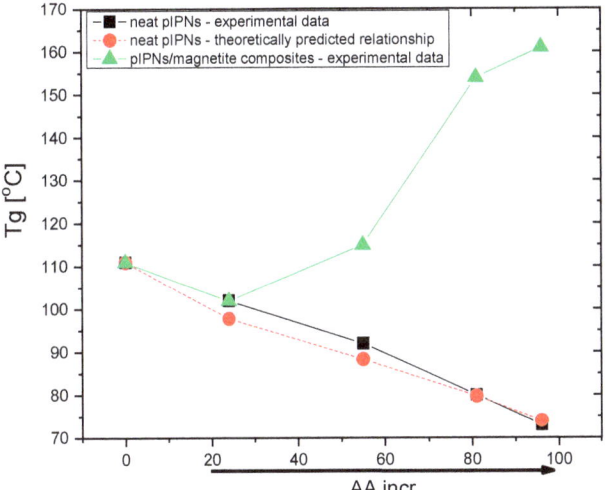

Figure 10. Dependence of pIPNs' Tg on the PAA content for: (■) neat pIPNs—experimental data; (•) neat pIPNs—theoretically predicted relationship using the additivity law; and (▲) pIPNs/magnetite composites.

In contrast, for pIPNs/magnetite composites, Tg increases as PAA content increases (Table S6, Figure 10). As the ATR-IR data have shown, there is an interaction between COO$^-$ anions and Fe^{n+} ions (Table S4), which imposes additional conformational constraints on the polymer segments involved in the binding of Fe ions. The reduction of the polymer flexibility is accompanied by an increase in Tg as reported for e.g., poly(ethylene oxide) interaction with Na$^+$ and Li$^+$ ions [28]. The authors report an exponential increase in Tg as salt concentration increases, which enhances the polymer–metal ion interactions. The case here is similar: the number of interactions between PAA and Fe ions increases as PAA content increases (Figure 10).

The Tg values for AA0X and AA20X do not show any deviation from the respective neat pIPNs, which could mean that, once formed, the magnetite does not strongly interact with the polymer network, which is also revealed by ATR-IR (Figure S5, Table S4). Thus, the positive deviation in the Tg dependence on PAA content is related mainly to the interaction between –COO$^-$ pendant groups and Fe ions [29].

2.9. Scanning Electron Microscopy

The morphology of the broken surface of two pIPNs/magnetite composites are presented in Figure 11. Both samples were chosen as they present two border cases: AA80X has a lower iron content than AA20X. Moreover, the latter is proven to contain magnetite rather than Fe ions (Figure 5). The in situ-formed magnetite particles in AA80X could be clearly seen as bright spots that are evenly dispersed within the polymer matrix (Figure 11A). The SEM image of the AA20X sample reveals a greater size of the formed magnetite particles (Figure 11B) when compared to the particles seen in AA80X. The average size of magnetite particles in AA80X is ~200 nm (Figure 11C) while in AA20X they have an average diameter of ~400 nm (Figure 11D). These results are well aligned with the XRD data and confirm the conclusion that the copolymer composition controls the process of in situ iron oxide formation. Moreover, the smaller mesh size of the pIPNs provides a greater saturation in its loops, leading to the formation of magnetite particles with higher crystallinity.

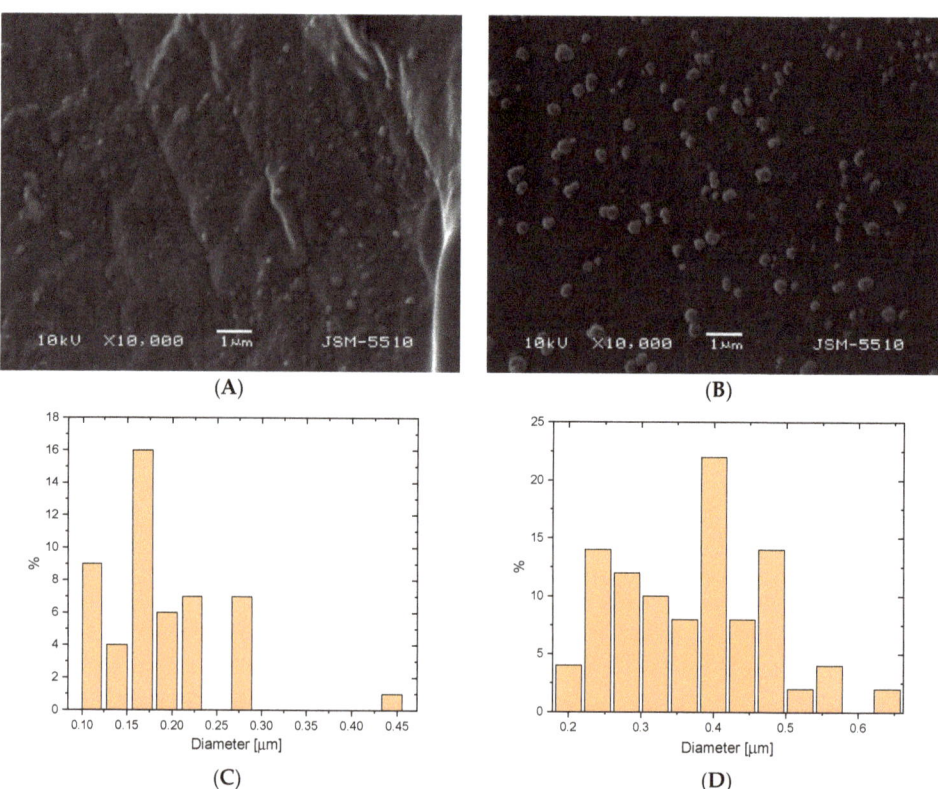

Figure 11. Morphology and magnetite particles' size distribution in pIPNs/magnetite composites for: AA80X (**A**) and (**C**); AA20X (**B**) and (**D**).

2.10. Magnetic Properties

The magnetic properties of pure magnetite (MPS) and the pIPNs/magnetite composites were studied using a vibrating sample magnetometer (VSM) at room temperature (Figure 12).

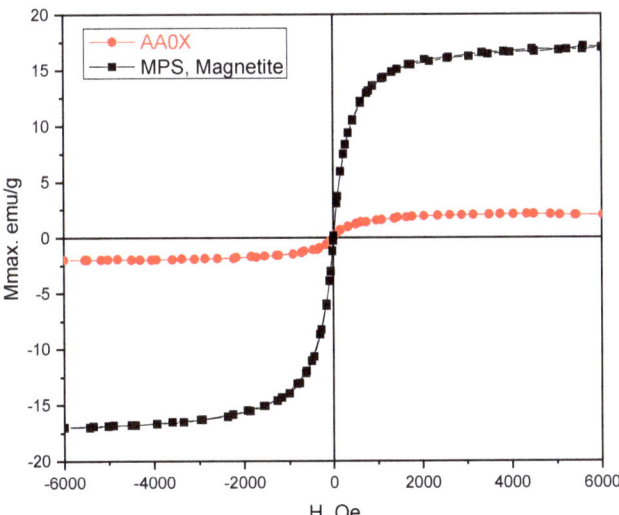

Figure 12. Magnetization of neat MPS and PAAM/PAAM pIPNs/magnetite composites.

The magnetization response, i.e., magnetization versus the applied magnetic field, has the well-known S-shaped curve, with coercive field and remanent magnetization very close (within the experimental error) to zero. Thus, both the pure magnetite (MPS) as well as magnetite/pIPNs composite hydrogels exhibit superparamagnetic behavior, i.e., they become magnetized when a magnetic field is applied but have no permanent magnetization (remanence) after the magnetic field is removed. It is clearly understood from the literature that magnetite nanoparticles with particle sizes of less than 20 nm are superparamagnetic and single domain, since 20 nm is less than the domain size for that material [30]. The magnetization values for iron oxide nanoparticles are usually between 30 and 50 emu/g [31], while higher values (e.g., 90 emu/g) have been observed for bulk materials. Here, the lower values for AA0X magnetization (~3 emu/g) could be related mainly to the small particle sizes (less than 20 nm), as confirmed by SEM (Figure 11). The factors which are known to contribute to the magnetization value of superparamagnetic iron oxide nanoparticles, are (i) the size of the particles; (ii) the spacing between the nanoparticles (coatings such as silica or polymers separate the magnetic domains, allowing each individual magnetite particle to act independently and thus enhancing the net magnetism per gram); and (iii) the crystalline structure of the iron oxide [31]. The size and the crystallinity are the most probable factors contributing to the lower magnetization of MPS as compared to that of similar particles reported in the literature.

The saturation magnetization of the hybrid hydrogel at room temperature is significantly lower than that of the pure magnetite samples. The cause is the diamagnetic contribution of the polymer matrix within which the magnetite nanoparticles are evenly dispersed, the mass of which was not possible to estimate and subtract.

3. Conclusions

This study reveals the synthesis of novel P(AA-co-AAM)/PAAM pIPNs and their composites with magnetite, obtained via in situ deposition. Formation of magnetite particles takes place via a complex mechanism involving the predominant role of PAA as an

Fe^{2+}/Fe^{3+} binding component. The results demonstrate that the pIPNs' composition is a powerful tool to control the size, crystallinity and quantity of the magnetite particles deposited in situ. SEM and XRD studies clearly confirm these conclusions. Moreover, the magnetite nanoparticles formed in situ act as additional crosslinks for the pIPNs networks, resulting in decreased swelling as well as reduced pH and ionic strength responsiveness of the pIPNs/magnetite composites when compared to the neat pIPNs. The valuation of the magnetic properties of the obtained composite hydrogels reveals their superparamagnetic properties with magnetization of ~3 emu/g. The results are encouraging and could be further used to widen the application of IPNs, in particular pIPNs, as templates for the preparation of polymer nanocomposites. Such novel inorganic/organic composites could find a wide range of applications as smart systems with defined pH and salt concentration responsiveness.

4. Materials and Methods

4.1. Materials

Iron (III) chloride, anhydrous and cyclohexane were purchased from Fisher Chemicals (Fisher Scientific, Waltham, MA, USA). Iron (II) sulfate heptahydrate, acrylic acid, acrylamide, potassium persulfate, N, N'-methylene-bis-acrylamide, N, N, N', N'-tetramethylethylene diamine (TEMED), sodium hydroxide, nitric acid, perchloric acid and phosphoric acid were purchased from Sigma Aldrich, USA. Polyacrylamide (PAAM) (Mw > 5,000,000) was purchased from BDH Laboratory reagents, Poole, England. All reagents were used as received.

4.2. Methods

4.2.1. Synthesis of P(AA-co-AAM)/PAAM pIPNs

P(AA-co-AAM)/PAAM pIPNs were synthesized by free radical polymerization. In summary, a defined amount of each monomer (Table 4) was dissolved in 1 wt% aqueous solution of linear PAAM (Mw = 5,000,000) giving a final total monomer concentration of 1.4 M. The initiator potassium persulfate (0.3 mol% to the total monomer content) and the crosslinking agent, N, N'-methylene-bis-acrylamide (2 mol% to the total monomer content) were added and the solution was homogenized using a magnetic stirrer at 350 rpm until all components were fully dissolved. Next, TEMED (0.2 v/v%) was added and the solution was further homogenized for 3 min. The obtained solution was placed between two glasses separated with a 1 mm thick rubber spacer and the polymerization took place over 45 min at 60 °C. After the end of the polymerization period, the obtained hydrogels were carefully removed from the glasses and placed in distilled water to purify them of any residuals. Water was changed 3 times daily until no residuals were found in the wastewater, as monitored by UV spectrophotometry. The purified hydrogels were cut into disk-shaped pieces and left to dry.

Table 4. Composition of the initial polymerization solutions as well as of the P(AA-co-AAM) copolymer (obtained by using the Mayo–Lewis equation).

Sample	Initial Polymerization Solution Composition			Reactivity Ratios Values		Copolymer Composition	
	AA [mol. %]	AAM [mol. %]	pH	r_1	r_2	AA * [mol. part]	AAM [mol. part]
AA100	100	0	2.7	PAA homopolymerization		0.96	-
AA80	80	20	2.9	1.34	0.69	0.81	0.19
AA50	50	50	3.3	1.34	0.69	0.55	0.45
AA20	20	80	3.7	1.28	0.82	0.24	0.76
AA0	0	100	6.2	PAAM homopolymerization		0	1

* N,N'-methylene-bis-acrylamide amount taken into account.

4.2.2. P(AA-co-AAM) Composition

The ratio between AA and AAM monomeric units in their copolymer was estimated using the Mayo–Lewis equation:

$$\frac{M^P_{AA}}{M^P_{AM}} = \frac{[AA] \cdot (r_1 \cdot [AA] + [AM])}{[AM] \cdot ([AA] + r_2 \cdot [AM])} \qquad (2)$$

where [AA] and [AM] are the molar concentrations of the monomers AA and AAM in the starting solution. M^P_{AA} and M^P_{AM} are the molar fractions of the corresponding monomeric units in the copolymer. r_1 and r_2 are, respectively, the reactivity ratios for the AA/AAM copolymerization. As the reactivity ratio depends on the pH of the reaction mixture, its values were taken at the closest point to the respective initial comonomer solution pH (Table 4) [32].

4.2.3. Preparation of pIPNs/Magnetite Composite Hydrogels

An aqueous solution of two Fe salts, namely $FeCl_3$ (0.30 M) and $FeSO_4 \cdot 7H_2O$ (0.15 M), was prepared and dry disk-shaped samples of pIPNs were immersed in it for 72 h. Fe^{2+}/Fe^{3+} loaded pIPN hydrogels were taken out of the solution, washed with distilled water, and placed in 6 M NaOH for 72 h. The same procedure was repeated using two other concentrations of the mixed Fe^{n+} salts solutions: 10 times diluted, i.e., $FeCl_3$ (0.03 M) and $FeSO_4 \cdot 7H_2O$ (0.015 M) (composite hydrogels designated with "Y"), and 100 times diluted, i.e., $FeCl_3$ (0.003 M) and $FeSO_4 \cdot 7H_2O$ (0.0015 M) (composite hydrogels designated with "Z"), respectively. The obtained composite hydrogels were washed with distilled water for 7 days, the water being changed twice daily. All hydrogels were left to dry in ambient conditions. The compositions of the obtained composites are presented in Table 2.

AA100Fe and AA0Fe samples (Table 5) were obtained to evaluate the swelling in Fe^{n+} mixed salt aqueous solution of the neat PAA and PAAM, respectively. Similarly, AA100N and AA0N samples (Table 5) were obtained to evaluate the swelling in NaOH aqueous solution of the neat PAA and PAAM samples. Moreover, for the sake of comparison, neat magnetite particles were synthesized via co-precipitation. In brief, the aqueous solution of $FeCl_3$ (0.30 M) and $FeSO_4 \cdot 7H_2O$ (0.15 M) was alkalized with 6 M NaOH to reach pH = 11. The obtained black precipitate was magnetically decanted, washed five times with distilled water and dried under ambient conditions. These particles are designated as MPS in Table 5. Their size was found to be 388 ± 143 nm via DLS (Figure S16, Supplementary Information).

Table 5. pIPN hydrogels with in situ-formed magnetite particles via co-precipitation.

Sample Designation	AA [mol. %]	AAM [mol. %]	Fe^{3+} [mol/L]	Fe^{2+} [mol/L]	6 M NaOH
AA100X	100	0	0.30	0.15	yes
AA80X	80	20	0.30	0.15	yes
AA50X	50	50	0.30	0.15	yes
AA20X	20	80	0.30	0.15	yes
AA0X	0	100	0.003	0.0015	yes
10 times dilution					
AA80Y	80	20	0.03	0.015	yes
AA50Y	50	50	0.03	0.015	yes
AA20Y	20	80	0.03	0.015	yes

Table 5. Cont.

Sample Designation	AA [mol. %]	AAM [mol. %]	Fe^{3+} [mol/L]	Fe^{2+} [mol/L]	6 M NaOH
		100 times dilution			
AA80Z	80	20	0.003	0.0015	yes
AA50Z	50	50	0.003	0.0015	yes
AA20Z	20	80	0.003	0.0015	yes
AA100Fe *	100	0	0.30	0.15	no
AA0Fe *	0	100	0.30	0.15	no
AA100N *	100	0	no	no	yes
AA0N *	0	100	no	no	yes
MPS *	-	-	0.003	0.0015	yes

* Samples obtained to be as referent.

4.2.4. Swelling Behavior

The volume of the dry polymer network, V_d, and the volume of the hydrogel at equilibrium swelling in water, V_s, were determined using the following equations:

$$V_d = \frac{m_{d,a} - m_{d,h}}{\rho} \tag{3}$$

$$V_s = \frac{m_{s,a} - m_{s,h}}{\rho} \tag{4}$$

Here $m_{d,a}$ and $m_{d,h}$ are the weights of the dry polymer networks measured in air and in cyclohexane, respectively, while $m_{s,a}$ and $m_{s,h}$ are the weights of the swollen hydrogels measured in air and in cyclohexane, respectively.

The polymer volume fraction, $v_{2,s}$, in the swollen state was calculated by using the equation:

$$v_{2,s} = \frac{V_d}{V_s} \tag{5}$$

The number average molecular mass between crosslinks ($\overline{M_c}$) was determined by the swelling experiments according to the Flory–Rehner equation:

$$\frac{1}{\overline{M_c}} = -\frac{\frac{\overline{v}}{V_1}[\ln(1 - v_{2,s}) + v_{2,s} + \chi v_{2,s}^2]}{(\sqrt[3]{v_{2,s}} - 0.5 v_{2,s})} \tag{6}$$

where V_1 is the molar volume of water (18 cm^3/mol) [21], $\overline{v}$ is the specific volume of the polymer (0.67) [33], and χ is the Flory–Huggins interaction parameter between the polymer and water, which is 0.5 for PAA and 0.49 for PAAM [33].

The theoretical number average molecular mass between crosslinks, $\overline{M_{c,t}}$, was calculated by using the equation:

$$\overline{M_{c,t}} = \frac{M_r}{2X} \tag{7}$$

where M_r is the molecular mass of the polymer repeat unit determined as weighted average and X is the nominal crosslinking ratio, calculated as the molar ratio between total monomer content and the crosslinking agent.

The network mesh size, ξ, was calculated using the equation:

$$\xi = \frac{1}{\sqrt[3]{v_{2,s}}} \sqrt{\frac{2 C_n \overline{M_c}}{M_r}} \, l \tag{8}$$

Here C_n is the Flory characteristic ratio (C_n^{PAA} = 6.7 and C_n^{PAAM} 2.72) [33,34], and l is the carbon–carbon bond length (1.54 Å).

4.2.5. Equilibrium Swelling Ratio (ESR)

Both the pIPNs and the pIPNs/magnetite composites were left to swell in water until they reached a constant weight. Next, their equilibrium swelling ratio (ESR) was determined by the equation:

$$\text{ESR} = \frac{m_{sw} - m_{dry}}{m_{dry}} \tag{9}$$

where m_{sw} and m_{dry} denote the weights of the sample in swollen state and dry state, respectively. pH of water used in the experiment was measured to be ~5.6 (Hanna instruments HI 2211 pH/ORP meter).

4.2.6. pH Responsive Behavior of pIPNs and pIPNs/Magnetite Hydrogels

The pH responsiveness of pIPNs and the pIPNs/magnetite composite hydrogels were determined in the pH range from 3 to 11 at 25 ± 1 °C. Briefly, dry disk-shaped samples (4.5 mm diameter) were swollen in buffer solution with defined pH until they reached a constant weight (for ~24 h). The swelling ratio at the respective pH (SR^{pH}) was calculated by using the equation:

$$SR^{pH} = \frac{m_{sw}^{pH} - m_{dry}}{m_{dry}} \tag{10}$$

where m_{sw}^{pH} and m_{dry} denote, respectively, the weights of the sample in its swollen state at certain pH and in its dry state. The buffer solutions were prepared following a procedure described in the literature [35].

4.2.7. Temperature Responsive Behavior of pIPNs and the pIPNs/Magnetite Composite Hydrogels

The temperature responsiveness of pIPNs and their magnetic composites was tested in the temperature range from 25 to 55 °C in water. Dry disk-shaped samples (4.5 mm diameter) were swollen in water at a defined temperature for 8 h. The swelling ratio at the defined temperature (SR^{Temp}) was calculated by using the equation:

$$SR^{Temp} = \frac{m_{sw}^{Temp} - m_{dry}}{m_{dry}} \tag{11}$$

where m_{sw}^{Temp} and m_{dry} denote the weight of the sample in its swollen state at certain temperature and in its dry state, respectively.

4.2.8. Ionic Strength Responsive Behavior of pIPNs and pIPNs/Magnetite Composite Hydrogels

The ionic strength responsiveness of pIPNs and the pIPNs/magnetite composites was measured in aqueous NaCl solutions with concentration ranging from 0.001 M to 1 M as well as in pure water (0 M). In brief, dry disk-shaped samples (4.5 mm diameter) were swollen in a solution with certain NaCl concentration until they reached a constant weight (~24 h). The swelling ratio (SR^I) was calculated for each NaCl concentration by using the equation:

$$SR^I = \frac{m_{sw}^I - m_{dry}}{m_{dry}} \tag{12}$$

where m_{sw}^I and m_{dry} denote the weights of the sample in its swollen state at certain NaCl concentration and in its dry state, respectively.

4.2.9. Attenuated Total Reflectance-FTIR (ATR-FTIR)

pIPNs and the pIPNs/magnetite composites were characterized using infrared spectroscopy (IR) in a regime of attenuated total reflectance by using an IRAffinity-1 Shimadzu Fourier transform infrared (FTIR) spectrophotometer with MIRacle attenuated total reflectance attachment. The samples were studied as dry solids, without any preliminary treatment.

4.2.10. Differential Scanning Calorimetry (DSC)

Differential scanning calorimetry tests were performed on DSC apparatus Q200, TA Instruments, USA. Dry samples (~5–6 mg) were heated from −50 to 150 °C, then cooled to −50 °C before being heated again to 230 °C at a 10 °C/min heating rate under nitrogen flow (50 mL/min). The results that were used and reported within the study are from the second heating run.

4.2.11. X-ray Diffraction (XRD)

A Siemens D500 diffractometer, Germany, with secondary monochromator and Cu-K$_\alpha$ radiation was used to obtain the diffractograms over 2θ range of 10–80° with a step of 0.03° and count time of 10 s. The search-match program Match3 was used to identify the crystal phases. The crystallite size D_{hkl} (in Å) in direction perpendicular to the (hkl) planes (311) and (400) was calculated according to Scherrer's formula:

$$D_{\langle hkl \rangle} = \frac{0.9 * \lambda}{\beta_{\frac{1}{2}} \cos \theta} \qquad (13)$$

where λ = 1.542 Å is the wavelength used, 2θ is the reflection position and $\beta_{1/2}$, in rad, is the physical integral width of the reflection hkl positioned at 2θ. Gaussian function was used to approximate the reflections. $\beta_{1/2}$ was obtained from the experimental width B and instrumental width b.

4.2.12. Iron Content Determination in pIPNs/Magnetite Composites
Flame Atomic Absorption Spectroscopy

Pieces from each pIPNs/magnetite composite were weighed (~0.2–0.3 g) and immersed in 2 mL conc. nitric acid in a 25 mL beaker. The beaker was covered with watch glass and then heated on a hot plate at 90–100 °C for 1 hour. This time was enough to partially degrade the polymer. The solution was then cooled down to room temperature. Then, 1 mL conc. perchloric acid was added and the beaker was heated again to 100–120 °C until white fumes appeared, and complete polymer degradation was achieved. After cooling down, the solution was diluted with distilled water and transferred to a 25 mL volumetric flask. The concentration of Fe ions was determined by flame atomic absorption spectroscopy (Perkin Elmer Analyst 400, Waltham, MA, USA) using a calibrating curve (Figure S17, Supplementary Information).

Scanning Electron Microscopy (SEM) with Energy-Dispersive X-ray Spectroscopy (EDX)

To determine the iron content as well as the magnetite particles' dispersion within the composites, the fractured surface of dry samples was covered with a thin carbon film (~10 nm). The samples were examined under a scanning electron microscope Lyra 3 XMU (Tescan), operating at 10 kV and coupled with EBSD and EDX analysis systems (Quantax 200, Bruker, Billerica, MA, USA).

Transmission Electron Microscopy (TEM)

Dry pIPNs/magnetite composites were cut into slices of ~100 nm width using an ultramicrotome (Leica EM UC7). The slices were examined under HR TEM (JEOL, JEM-2100), operating at 200 kV.

Magnetic Measurements

Magnetization curves were measured at room temperature using a vibrating sample magnetometer (VSM) in fields up to 6 kOe. The samples were prepared by pressing the powder into cylindrical ($\varnothing$ = 3 mm, h = 10 mm) quartz containers so that the particles were unable to move during the measurements.

Supplementary Materials: The following supporting information can be downloaded at: https://www.mdpi.com/article/10.3390/gels9050365/s1, [12,36–41] Table S1. ANOVA data for the dependence of ESR on the pIPNs compostion for neat IPNs, Table S2. An effect size of groups' ESR. Generalized eta squared-ges of 0.945 (95%) means that 95% of the change in the ESR can be accounted for the treatment conditions. Figure S1. One-way single factor ANOVA results. Variance of ESR between groups. Figure S2. P.adjust dependence from Tukey test for significance. Figure S3. ATR-IR spectrum of neat pIPNs (A) and samples AA0 and AA20, showing the band shifting due to hydrogen bonding (B). Figure S4. ATR-IR spectra of the pIPNs, obtained with PAA homopolymer (AA100N) and PAAM hopomopolymer (AA0N) after treatment with 6 M NaOH for 72 h. Table S3. Band assignments the pIPNS IR spectra with homopolymer PAAM (AA0 sample). Table S4. Band assignments of the pIPNs with homopolymer PAA (AA100 sample). Figure S5. ATR-IR spectra of pIPNs (A) with homopolymer PAAM: neat (AA0); with absorbed Fe ions (AA0Fe); treated with 6M NaOH for 72 h (AA0N); and the respective magnetite/PAAM composite AA0X; and (B) with homopolymer PAA: neat (AA100); with absorbed Fe ions (AA100Fe); treated with 6M NaOH for 72 h (AA100N); and the respective magnetite/pIPNs composite (AA100X). Figure S6. Equilibrium swelling ratio of pIPNs hydrogels with in situ formed iron oxide obtained via using different Fe aqueous solutions concentration: X (Fe^{3+}—0.30 mol/L; Fe^{2+}—0.15 mol/L); Y (Fe^{3+}—0.030 mol/L; Fe^{2+}—0.015 mol/L) Z (Fe^{3+}—0.003 mol/L; Fe^{2+}—0.0015 mol/L). Table S5. ANOVA analysis on the dependence of initial iron solution on the ESR of the resulting hybrid hydrogels. Figure S7. Iron content as determined by Flame Atomic Absorption Spectroscopy (FAAS) in pIPNs composite materials with in situ formed iron oxide obtained via using different Fe aqueous solutions concentration: X (Fe^{3+}—0.30 mol/L; Fe^{2+}—0.15 mol/L); Y (Fe^{3+}—0.030 mol/L; Fe^{2+}—0.015 mol/L) Z (Fe^{3+}—0.003 mol/L; Fe^{2+}—0.0015 mol/L). Figure S8. XRD patterns of sample AA20X in is dry and swollen state. Figure S9. XRD diffractograms of pIPNs/magnetite composites with different composition with in situ formed magnetite. Figure S10. Comparison between the swelling ratios of pIPNs and the pIPNs/magnetite composites at pH = 5. Figure S11. Temperature responsiveness of pIPNs/magnetite composite hydrogels. Figure S12. DSC thermograms of neat pIPNs (the 2nd heat scan from heat/cool/heat run is presented). Figure S13. DSC thermograms of pIPNs/magnetite composites (the 2nd heat scan from the heat/cool/heat run is presented). Table S6. Experimentally determined Tg of pIPNs and pIPNs/magnetite composites. Figure S14. ATR-IR spectrum of pIPNs/magnetite composites. It is clearly visible there is no significant difference between samples. Figure S15. Normalized concentration of TC removed by AA0X (A) and by of potassium persulfate activation (B). Figure S16. Size of the neat magnetite particles MPS as revealed by DLS. Figure S17. Calibration curve for Fe^{n+} ions, used in FAAS experiments.

Author Contributions: Conceptualization: M.S. and E.V.; methodology: M.S., M.G. and A.A.A.; data curation: M.S., D.T. and A.A.A.; writing: M.S., M.G. and E.V.; original draft preparation: M.S. and E.V.; project administration: M.S. All authors have read and agreed to the published version of the manuscript.

Funding: The financial support from BNSF under grant KΠ-06-M59/4//19.11.2021 is gratefully acknowledged.

Institutional Review Board Statement: Not applicable.

Informed Consent Statement: Not applicable.

Data Availability Statement: The raw/processed data required to reproduce these findings cannot be shared at this time as the data also form part of an ongoing study.

Acknowledgments: The authors acknowledge the support of Stefan Tsakovski (Faculty of Chemistry and Pharmacy, Sofia University) for the statistical analysis of data and of Martin Tsvetkov (Laboratory of Chemistry of Rare and Rare Earth Elements, Faculty of Chemistry and Pharmacy, Sofia University) for sorption experiments.

Conflicts of Interest: The authors declare no conflict of interest.

References

1. Dragan, E.S. Design and applications of interpenetrating polymer network hydrogels: A review. *Chem. Eng. J.* **2014**, *243*, 572–590. [CrossRef]
2. Tavakoli, J.; Tang, Y. Hydrogel Based Sensors for Biomedical Applications: An Updated Review. *Polymers* **2017**, *9*, 364. [CrossRef]
3. Mantha, S.; Pillai, S.; Khayambashi, P.; Upadhyay, A.; Zhang, Y.; Tao, O.; Pham, H.M.; Tran, S.D. Smart Hydrogels in Tissue Engineering and Regenerative Medicine. *Materials* **2019**, *12*, 3323. [CrossRef]
4. Macnaught, A.D.; Wilkinson, A.; Union, I. *Compendium of Chemical Terminology: IUPAC Recommendations*; Oxford Blackwell Science: Hoboken, NJ, USA, 1997.
5. Sperling, L.H. *Polymeric Multicomponent Materials: An Introduction*, 2nd ed.; John Wiley & Sons: Nashville, TN, USA, 1997.
6. Katz, E.; Pita, M. Biomedical applications of magnetic particles. In *Fine Particles in Medicine and Pharmacy*; Springer: Boston, MA, USA, 2012; pp. 147–173.
7. Sivudu, K.S.; Rhee, K.Y. Preparation and Characterization of PH-Responsive Hydrogel Magnetite Nanocomposite. *Colloids Surf. A Physicochem. Eng. Asp.* **2009**, *349*, 29–34. [CrossRef]
8. Sanchez, L.M.; Martin, D.A.; Alvarez, V.A.; Gonzalez, J.S. Polyacrylic Acid-Coated Iron Oxide Magnetic Nanoparticles: The Polymer Molecular Weight Influence. *Colloids Surf. Physicochem. Eng. Asp.* **2018**, *543*, 28–37. [CrossRef]
9. Xie, S.; Zhang, B.; Wang, L.; Wang, J.; Li, X.; Yang, G.; Gao, F. Superparamagnetic iron oxide nanoparticles coated with different polymers and their MRI contrast effects in the mouse brains. *Appl. Surf. Sci.* **2015**, *326*, 32–38. [CrossRef]
10. Dheyab, M.A.; Aziz, A.A.; Jameel, M.S.; Noqta, O.A.; Khaniabadi, P.M.; Mehrdel, B. Simple rapid stabilization method through citric acid modification for magnetite nanoparticles. *Sci. Rep.* **2020**, *10*, 10793.
11. Shah, S.T.; Yehya, W.A.; Saad, O.; Simarani, K.; Chowdhury, Z.; Alhadi, A.A.; Al-Ani, L.A. Surface Functionalization of Iron Oxide Nanoparticles with Gallic Acid as Potential Antioxidant and Antimicrobial Agents. *Nanomaterials* **2017**, *7*, 306. [CrossRef] [PubMed]
12. Namdeo, M.; Bajpai, S.K.; Kakkar, S. Preparation of a magnetic-field-sensitive hydrogel and preliminary study of its drug release behavior. *J. Biomater. Sci. Polym. Ed.* **2009**, *20*, 1747–1761. [CrossRef]
13. Starodubtsev, S.G.; Saenko, E.V.; Dokukin, M.E.; Aksenov, V.L.; Klechkovskaya, V.V.; Zanaveskina, I.S.; Khokhlov, A.R. Formation of Magnetite Nanoparticles in Poly(Acrylamide) Gels. *J. Phys. Condens. Matter* **2005**, *17*, 1471–1480. [CrossRef]
14. Arakaki, A.; Nakazawa, H.; Nemoto, M.; Mori, T.; Matsunaga, T. Formation of magnetite by bacteria and its application. *J. R. Soc. Interface* **2008**, *5*, 977–999. [CrossRef]
15. Chełminiak, D.; Ziegler-Borowska, M.; Kaczmarek, H. Synthesis of magnetite nanoparticles coated with poly(acrylic acid) by photopolymerization. *Mater. Lett.* **2016**, *164*, 464–467. [CrossRef]
16. Drozdov, A.S.; Prilepskii, A.Y.; Koltsova, E.M.; Anastasova, E.I.; Vinogradov, V.V. Magnetic Polyelectrolyte-Based Composites with Dual Anticoagulant and Thrombolytic Properties: Towards Optimal Composition. *J. Sol-Gel Sci. Technol.* **2020**, *95*, 771–782. [CrossRef]
17. Chou, F.-Y.; Lai, J.-Y.; Shih, C.-M.; Tsai, M.-C.; Lue, S.J. In Vitro Biocompatibility of Magnetic Thermo-Responsive Nanohydrogel Particles of Poly(N-Isopropylacrylamide-Co-Acrylic Acid) with Fe_3O_4 Cores: Effect of Particle Size and Chemical Composition. *Colloids Surf. B Biointerfaces* **2013**, *104*, 66–74. [CrossRef]
18. Hosny, N.M.; Abbass, M.; Ismail, F.; El-Din, H.M.N. Radiation Synthesis and Anticancer Drug Delivery of Poly(Acrylic Acid/Acrylamide) Magnetite Hydrogel. *Polym. Bull.* **2022**, *80*, 4573–4588. [CrossRef]
19. Edwards, M.; Benjamin, M.M.; Ryan, J.N. Role of organic acidity in sorption of natural organic matter (NOM) to oxide surfaces. Colloids Surf. *Physicochem. Eng. Asp.* **1996**, *107*, 297–307. [CrossRef]
20. Todica, M.; Pop, C.V.; Udrescu, L.; Stefan, T. Spectroscopy of a Gamma Irradiated Poly(Acrylic Acid)-Clotrimazole System. *Chin. Phys. Lett.* **2011**, *28*, 128201. [CrossRef]
21. Gudeman, L.F.; Peppas, N.A. Preparation and Characterization of PH-Sensitive, Interpenetrating Networks of Poly(Vinyl Alcohol) and Poly(Acrylic Acid). *J. Appl. Polym. Sci.* **1995**, *55*, 919–928. [CrossRef]
22. Andrews, R.D.; Tobolsky, A.V.; Hanson, E.E. The Theory of Permanent Set at Elevated Temperatures in Natural and Synthetic Rubber Vulcanizates. *Rubber Chem. Technol.* **1946**, *19*, 1099–1112. [CrossRef]
23. El-Zahhar, A.A.; Ashraf, I.M.; Idris, A.M.; Zkria, A. Pronounced Effect of PbI_2 Nanoparticles Doping on Optoelectronic Properties of PVA Films for Photo-Electronic Applications. *Phys. B Condens. Matter* **2022**, *630*, 413604. [CrossRef]
24. Mayer, C.R.; Cabuil, V.; Lalot, T.; Thouvenot, R. Magnetic Nanoparticles Trapped in PH 7 Hydrogels as a Tool to Characterize the Properties of the Polymeric Network. *Adv. Mater.* **2000**, *12*, 417–420. [CrossRef]
25. Michaels, A.S.; Morelos, O. Polyelectrolyte Adsorption by Kaolinite. *Ind. Eng. Chem.* **1955**, *47*, 1801–1809. [CrossRef]
26. Chun, M.-K.; Cho, C.-S.; Choi, H.-K. Characteristics of Poly(Vinyl Pyrrolidone)/Poly(Acrylic Acid) Interpolymer Complex Prepared by Template Polymerization of Acrylic Acid: Effect of Reaction Solvent and Molecular Weight of Template. *J. Appl. Polym. Sci.* **2004**, *94*, 2390–2394. [CrossRef]
27. Li, K.; Chen, K.; Wang, Q.; Zhang, Y.; Gan, W. Synthesis of Poly(Acrylic Acid) Coated Magnetic Nanospheres via a Multiple Polymerization Route. *R. Soc. Open Sci.* **2019**, *6*, 190141. [CrossRef] [PubMed]

28. Stolwijk, N.A.; Heddier, C.; Reschke, M.; Wiencierz, M.; Bokeloh, J.; Wilde, G. Salt-Concentration Dependence of the Glass Transition Temperature in PEO–NaI and PEO–LiTFSI Polymer Electrolytes. *Macromolecules* **2013**, *46*, 8580–8588. [CrossRef]
29. Salminen, L.; Karjalainen, E.; Aseyev, V.; Tenhu, H. Tough Materials through Ionic Interactions. *Front. Chem.* **2021**, *9*, 721656. [CrossRef]
30. Namanga, J.; Foba, J.; Ndinteh, D.T.; Yufanyi, D.M.; Krause, R.W.M. Synthesis and Magnetic Properties of a Superparamagnetic Nanocomposite "Pectin-Magnetite Nanocomposite". *J. Nanomater.* **2013**, *2013*, 137275. [CrossRef]
31. Kucheryavy, P.; He, J.; John, V.T.; Maharjan, P.; Spinu, L.; Goloverda, G.Z.; Kolesnichenko, V.L. Superparamagnetic Iron Oxide Nanoparticles with Variable Size and an Iron Oxidation State as Prospective Imaging Agents. *Langmuir* **2013**, *29*, 710–716. [CrossRef] [PubMed]
32. Rintoul, I.; Wandrey, C. Polymerization of Ionic Monomers in Polar Solvents: Kinetics and Mechanism of the Free Radical Copolymerization of Acrylamide/Acrylic Acid. *Polymer* **2005**, *46*, 4525–4532. [CrossRef]
33. Thakur, A.; Wanchoo, R.K.; Singh, P. Structural Parameters and Swelling Behavior of PH Sensitive Poly(Acrylamide-Co-Acrylic Acid) Hydrogels. *Chem. Biochem. Eng. Q.* **2011**, *25*, 181–194.
34. Finch, C.A. *Polymer Handbook*, 3rd ed.; Brandrup, J., Immergut, E.H., Eds.; British Polymer Journal; Wiley-Interscience: Chichester, UK, 1990; Volume 23, p. 277.
35. Pourjavadi, A.; Mahdavınıa, G.R. Superabsorbency, pH-Sensitivity and Swelling Kinetics of Partially Hydrolyzed Chitosan-g-poly(Acrylamide) Hydrogels. *Turk. J. Chem.* **2006**, *30*, 595–608.
36. Lindén, L.; Rabek, J.F. Structures and Mechanisms of Formation of Poly(Acrylic Acid)-Iron (II and III) Chloride Gels in Water and Hydrogen Peroxide. *J. Appl. Polym. Sci.* **1993**, *50*, 1331–1341. [CrossRef]
37. Filho, E.; Brito, E.; Silva, R.; Streck, L.; Bohn, F.; Fonseca, J. Superparamagnetic Polyacrylamide/Magnetite Composite Gels. *J. Dispers. Sci. Technol.* **2021**, *42*, 1504–1512. [CrossRef]
38. Alvarez-Gayosso, C.A.; Canseco, M.; Estrada, R.; Palacios-Alquisira, J.; Hinojosa, J.; Castano, V. Preparation and Micro-structure of Cobalt(Iii) Poly (Acrylate) Hybrid Materials. *Int. J. Basic Appl. Sci.* **2015**, *4*, 255. [CrossRef]
39. Leung, W.M.; Axelson, D.E.; Van Dyke, J.D. Thermal Degradation of Polyacrylamide and Poly(Acrylamide-Co-Acrylate). *J. Polym. Sci. Part A Polym. Chem.* **1987**, *25*, 25–1825. [CrossRef]
40. Simeonov, M.S.; Apostolov, A.A.; Vassileva, E.D. In Situ Calcium Phosphate Deposition in Hydrogels of Poly(Acrylic Acid)–Polyacrylamide Interpenetrating Polymer Networks. *RSC Adv.* **2016**, *6*, 16274–16284. [CrossRef]
41. Lee, D.; Kim, S.; Tang, K.; De Volder, M.; Hwang, Y. Oxidative Degradation of Tetracycline by Magnetite and Persulfate: Performance, Water Matrix Effect, and Reaction Mechanism. *Nanomater* **2021**, *11*, 2292. [CrossRef]

Disclaimer/Publisher's Note: The statements, opinions and data contained in all publications are solely those of the individual author(s) and contributor(s) and not of MDPI and/or the editor(s). MDPI and/or the editor(s) disclaim responsibility for any injury to people or property resulting from any ideas, methods, instructions or products referred to in the content.

Article

Tough, Injectable Calcium Phosphate Cement Based Composite Hydrogels to Promote Osteogenesis

Yazhou Wang [1,2,†], Zhiwei Peng [3,†], Dong Zhang [4,*] and Dianwen Song [1,5,*]

1. Department of Orthopedics, Shanghai General Hospital of Nanjing Medical University, Shanghai 201600, China
2. Department of Orthopedics, Shanghai Songjiang District Central Hospital, Shanghai 201620, China
3. Department of Orthopedics, The Second Affiliated Hospital of Xuzhou Medical University, Xuzhou 221000, China
4. The Wallace H. Coulter Department of Biomedical Engineering, Georgia Institute of Technology and Emory University, Atlanta, GA 30332, USA
5. School of Medicine, Shanghai Jiaotong University, Shanghai 200240, China
* Correspondence: dzhang470@gatech.edu (D.Z.); dianwen_song@126.com (D.S.)
† These authors contributed equally to this work.

Abstract: Osteoporosis is one of the most disabling consequences of aging, and osteoporotic fractures and a higher risk of subsequent fractures lead to substantial disability and deaths, indicating that both local fracture healing and early anti-osteoporosis therapy are of great significance. However, combining simple clinically approved materials to achieve good injection and subsequent molding and provide good mechanical support remains a challenge. To meet this challenge, bioinspired by natural bone components, we develop appropriate interactions between inorganic biological scaffolds and organic osteogenic molecules, achieving a tough hydrogel that is both firmly loaded with calcium phosphate cement (CPC) and injectable. Here, the inorganic component CPC composed of biomimetic bone composition and the organic precursor, incorporating gelatin methacryloyl (GelMA) and N-Hydroxyethyl acrylamide (HEAA), endow the system with fast polymerization and crosslinking through ultraviolet (UV) photo-initiation. The GelMA-poly (N-Hydroxyethyl acrylamide) (GelMA-PHEAA) chemical and physical network formed in situ enhances the mechanical performances and maintains the bioactive characteristics of CPC. This tough biomimetic hydrogel combined with bioactive CPC is a new promising candidate for a commercial clinical material to help patients to survive osteoporotic fracture.

Keywords: injectable hydrogels; osteogenesis; bone cement; biocompatible polymers

Citation: Wang, Y.; Peng, Z.; Zhang, D.; Song, D. Tough, Injectable Calcium Phosphate Cement Based Composite Hydrogels to Promote Osteogenesis. *Gels* **2023**, *9*, 302. https://doi.org/10.3390/gels9040302

Academic Editor: Richard J. Williams

Received: 6 March 2023
Revised: 29 March 2023
Accepted: 30 March 2023
Published: 3 April 2023

Copyright: © 2023 by the authors. Licensee MDPI, Basel, Switzerland. This article is an open access article distributed under the terms and conditions of the Creative Commons Attribution (CC BY) license (https://creativecommons.org/licenses/by/4.0/).

1. Introduction

Osteoporosis, a major worldwide health problem, is associated with substantial social, economic, and public health burdens. By 2030, approximately 13.3 million individuals in the United States older than 50 years are expected to have osteoporosis [1]. Fractures, the most important consequence of osteoporosis, are associated with enormous costs and substantial morbidity and mortality [2,3]. Roughly 9 million osteoporotic fractures occur worldwide each year [4], and approximately one in three women and one in five men aged 50 years or older will have a fragility fracture during their remaining lifetime. Furthermore, a total of 23% of the subsequent fractures occur within 1 year after the first fracture, and 54.3% occur within 5 years [5], indicating the treatment for the first fracture with an internal fixation system or bone cement alone is deemed insufficient, resulting in an urgent need for early anti-osteoporosis therapy after a first fracture to prevent subsequent fractures. However, achieving rapid recovery of local fractures to avoid long-term bedrest that can lead to further systemic osteoporosis, and at the same time, improving the total bone mass to avoid secondary fractures is still a great challenge.

Osteoporosis is a systemic skeletal disease characterized by reduced bone mass and microarchitectural deterioration of bone tissue leading to an increased risk of fragility fracture [3]. Osteoporosis is a chronic disease and long-term management is required. The purpose of treating patients with osteoporosis medication is to reduce the risk of fracture and subsequent pain and disability [6]. At present, most existing therapeutics used in the treatment of osteoporosis are anti-resorptive drugs, such as bisphosphonates, and bone anabolic agents, including denosumab and teriparatide [3,6]. If individuals at high risk of fractures do not receive appropriate treatment, this may result in further consequences. Vertebral fractures are the most common among osteoporotic fractures and due to poor bone quality, screw loosening and pull-out occur frequently in older osteoporotic patients, which presents several challenges to spine surgeons [7,8].

Common strategies for improving osteointegration include aesthetic contouring at the physical level and osteogenesis at the biochemical level [9–11]. Percutaneous vertebroplasty (PVP) and percutaneous kyphoplasty (PKP) are widely used in the treatment of osteoporotic vertebral compression fractures (OVCFs). PVP and PKP refer to a minimally invasive spine surgery technique that injects bone cement into the vertebral body through the pedicle or beside the pedicle to relieve back pain, increase the stability of the vertebral body, and restore the height of the vertebral body [12]. Pedicle screw fixation is widely used to treat spinal disease, and the number of spine surgeries in elderly patients with osteoporosis continues to increase worldwide due to the increasingly aged population [13,14]. To improve the pull-out strength of screws in the osteoporotic spine and decrease the risk of screw loosening, several techniques are used, such as using an expandable screw, enlarging the length and diameter of the screw, and using a cement-augmented pedicle screw (CAPS). Among these approaches, CAPS has been proven to be the most effective strategy for enhancing the fixation strength to improve pedicle screw stability in patients with osteoporosis [13,14]. However, combining simple clinically approved materials to achieve good injection and subsequent molding and provide good mechanical support remains a challenge. The common types of bone cement used clinically include polymethylmethacrylate (PMMA), calcium phosphate cement (CPC), calcium sulfate cement (CSC), and composite bone cement. Currently, PMMA bone cement is the most commonly used bone cement in PVP/PKP and CAPS, having advantages such as biocompatibility, injectability, and good mechanical properties [12,15]. However, PMMA also has various disadvantages, such as it cannot be degraded, a lack of biocompatibility, a propensity to cause surrounding tissue damage due to polymerization exotherm, and residual monomer toxicity [12]. In addition, the injection of PMMA bone cement into a vertebral body increases the possibility of fracture of the adjacent vertebral body.

The ideal bone cement is biocompatible, resorbable, osteoconductive, osteoinductive, and mechanically similar to bone. The study and development of new bone cement alternatives to PMMA is the focus of intensive investigations worldwide. In the last few decades, injectable hydrogels have gained increasing attention due to their structural similarities with the extracellular matrix, easy process conditions, and potential applications in minimally invasive surgery [15]. CPC, which has good osteoconductive and biocompatible capacity, presents an advantageous alternative material. Fortunately, with the rapid development in nanotechnology, nanomaterials are easily characterized (such as using X-ray and neutron diffraction to detect structure) [16–18] and evaluated (surface morphology and surface energy determined by atomic force microscopy) [19]. Thus, nanomaterials have been widely applied in the ferrimagnetic and optical domains [20,21]. The controllable preparation of CPC nano powder-formed CPC scaffolds has been widely applied in clinical application [22–25]. However, the time-consumer curing process, intrinsic uncontrolled brittleness, and poor washout resistance have limited its further integrated applications. To address these drawbacks of pure CPC, one of the major strategies is to integrate organic-inorganic phases and simulation tissue composite. Some peptide-based matrices endow materials with bioactive and mechanical properties, but they are limited by stringent synthesis processes [26,27]. Gelatin methacryloyl (GelMA) derived from collagen with injectable, bioactivity, and fast crosslinking progress has been widely studied [28–32]. However, its poor mechanical behavior has confined it to accelerating bone reconstruction. Introducing poly(ethylene glycol) diacrylate (PEGDA) to form GelMA/PEGDA hydrogel showed guided bone regeneration, however, the mechanical properties were still weak [28]. Thus, developing injectable, osteoblast-active, and tough materials is still a big challenge. In this study, we design an organic-inorganic precursor containing CPC, GelMA, and N-Hydroxyethyl acrylamide (HEAA) with fast gelation behavior [33–37]. The CPC and GelMA are supported to promote bone regeneration. The in situ-formed chemical and physical GelMA-poly(N-Hydroxyethyl acrylamide) (GelMA-PHEAA) network endows the composite with a uniquely tough structure that enhances the mechanical performance and maintains the bioactive characteristics of CPC. Due to the above properties, the resultant GelMA-PHEAA/CPC hydrogels impart superior tough mechanical properties and strong osteogenic ability.

Herein, bioinspired by natural bone structure, we develop a biomimetic bone structure that fully considers the need for appropriate interactions between inorganic osteogenic teriparatide and organic powerful biological scaffolds, achieving a scaffold that is both firmly loaded with CPC and able to provide strong mechanical support. The HEAA bridges in the system make the whole hydrogel network very tough, realizing the great storage of CPC and excellent osteogenic properties. Meanwhile, the components are Food and Drug Administration (FDA)-approved and well-suited to clinical translation. In summary, this bioactive injectable hydrogel is a novel promising therapy for fracture patients and well-suited to clinical commercialization.

2. Results and Discussion
2.1. Synthesis and Characterization of GelMA-PHEAA/CPC Hydrogels

In this study, a new injectable and bioactive hydrogel was designed. The schematic in Figure 1A illustrates the hydrogel preparation process. By introducing GelMA and HEAA monomer into the CPC precursor, a fast cross-linked homogeneous hydrogel system could be fabricated. Such a protocol allowed the in-situ gelation of hydrogels at localized defects and accelerated bone regeneration. As shown in Figure 1B, the typical vial inversion test proved that GelMA-PHEAA/CPC hydrogel could be easily formed under UV photo initiation in 2 min. The injection process in Figure 1C demonstrated that the precursor was able to plastically mold in a preset shape with fast cross-linking. Figure 1C(I–IV) show the precursor injection, defect filling, formation in situ, and final shape, respectively. These properties will allow the repairing of irregular defects in clinical applications. A rheological test was used to evaluate the processability of the used materials. Although the pre-solution contains macromolecule, monomer, and inorganic particles, the system still exhibits an obvious shear thinning phenomenon (Figure 1D). Due to the fast cross-linking ability, the system could form a stable network consisting of an organic-inorganic hybrid structure. A rheological frequency-sweep test (Figure 1E) showed that the storage modulus (G') of the GelMA-PHEAA/CPC precursor was lower than the loss modulus (G'') at the sol stage, and after polymerization, the G' of the GelMA-PHEAA/CPC hydrogels was always higher than the G'', which indicated that the hydrogels had both a stable structure and elasticity at a wide range of frequencies.

Figure 2A shows the SEM micrographs of the hydrogels. The pure GelMA hydrogel exhibited a smooth surface with few porous structures, while the GelMA-PHEAA hydrogel presented a flat smooth surface. However, the introduction of CPC in the GelMA system caused a remarkable increase in porosity. The CPC powder was loose within the GelMA networks. In addition, compared with the GelMA/CPC hydrogel, the GelMA-PHEAA/CPC hydrogel revealed a tighter structure, which allowed CPC to be well-dispersed in the system. The surface element detection in Figure 2B also illustrates that the GelMA-PHEAA/CPC hydrogel contains abundant bioactive particles with a homogeneous composition. The chemical structures of the precursor and hydrogels were tested using FT-IR spectroscopy (Figure 2C). The characteristic absorption peaks at 3300~3000 cm^{-1} of HEAA and the

GelMA macromolecule represented the unsaturated C-H vibration. Absorption peaks at 1680~1620 cm^{-1} represent the -C=C- vibration of HEAA, which was observed in both HEAA and the GelMA-PHEAA/CPC precursor. However, in the case of GelMA, this double bond vibration may overlap with the -NH- vibration whose characteristic peak lay at 1650~1500 cm^{-1}. After photo-initiation for 2 min, the double bond and unsaturated C-H vibration peaks in both GelMA-PHEAA and GelMA-PHEAA/CPC groups disappeared, indicating that both systems experienced in situ completed polymerization. In general, the complete polymerization of hydrogel precursors always results in less cytotoxicity for clinical application compared to the toxic monomers.

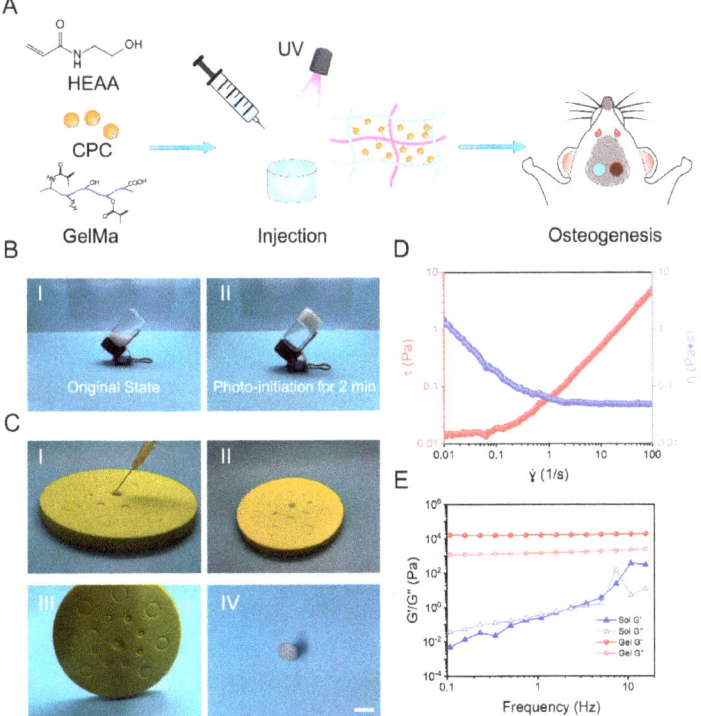

Figure 1. Schematic representations of GelMA-PHEAA/CPC hydrogels for bone regeneration (**A**). The optical images present the sol−gel transformation of hydrogels, and I,II represent the original state and after photo-initiation of GelMA-PHEAA/CPC precursor and GelMA-PHEAA/CPC hydrogel, respectively (**B**). The optical images present the injectability and ability to form in situ of the materials, and I–IV represent precursor injection, defect filling, formation in situ, and final shape, respectively (**C**). Rheological behavior of pre-solution and hydrogels under rotation ramp mode (**D**) and dynamic frequency sweep tests (**E**).

Next, to determine the phase composition of CPC in the composite GelMA-PHEAA/CPC hydrogels, X-ray diffraction (XRD) analysis was performed (Figure 2D). The CPC consisted of tetracalcium phosphate (TTCP, $Ca_4(PO_4)_2O$) and dicalcium phosphate anhydrous (DCPA, $CaHPO_4$), which could form hydroxyapatite (Hap) ($Ca_{10}(PO_4)_6(OH)_2$) in situ [38]. The broad peak at around $2\theta = 23°$ represented the polymer chain segment of the organic composite. Both samples exhibited typical peaks of HAp and anhydrate TTCP phase. However, the intensity of HAp and TTCP in GelMA-PHEAA/CPC was obviously weaker than CPC, which might be due to the shielding effect of the organic phase. Appropriate mechanical properties are vital for materials applied in bone defects. As mentioned, GelMA-

PHEAA/CPC hydrogel could be injectable with fast cross-linking, but the form of the organic-inorganic composite needs to be strong enough to support tissue regeneration. The compress-strain curves in Figure 2E,F show a comparison of the mechanical performance of each sample. The clinically used CPC was brittle with less than 10% compressibility, which is a major disadvantage in practical application. At the same time, while, due to its bioactivity and biocompatibility, GelMA is wildly researched in tissue regeneration, the soft and weak networks it forms limited its application in hard tissue repair. The introduction of CPC in GelMA would weaken the network. As shown in Figure 2F, the modulus of GelMA/CPC is one-third that of GelMA. However, by introducing PHEAA into the system, both Young's modulus and the compressibility of the materials were significantly promoted. GelMA/CPC and GelMA-PHEAA/CPC both showed lower mechanical properties than their hydrogel matrix. This may be attributed to the weak interaction between CPC dispersed in the system and the polymer, which decreases the strength of the polymer network connections. However, with the increase in compression deformation, the breaking stress of GelMA/CPC and GelMA-PHEAA/CPC became greater than that of their hydrogel matrix. These phenomena may be due to the strong interaction between CPC and the polymer network. In addition, the fluctuations in the curve of the GelMA-PHEAA/CPC hydrogel after 78% deformation indicated the local failure of the system; however, the materials could maintain their structural integrity at 90% strain with over 3.5 MPa stress. This property allowed the GelMA-PHEAA/CPC hydrogel to support a hard tissue structure and would reduce the potential risk of implant material rupture.

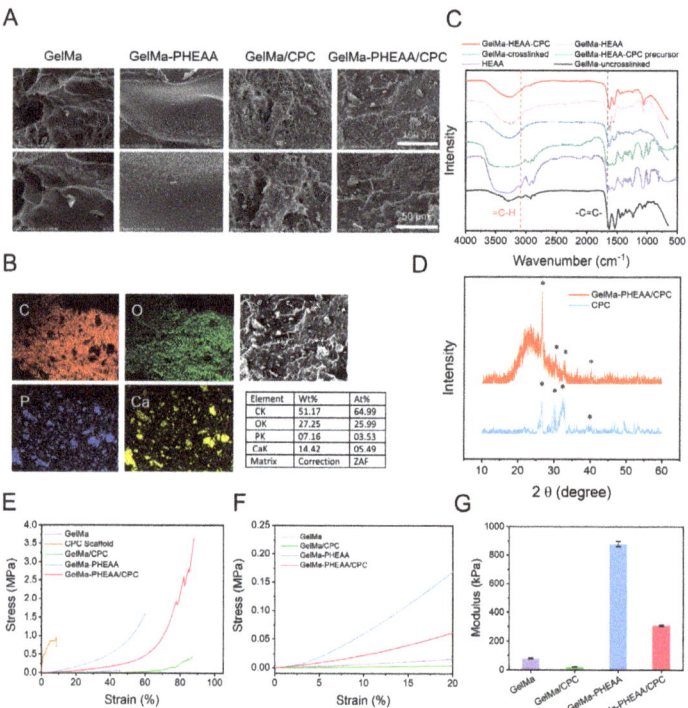

Figure 2. The characteristics of the hydrogels. (**A**) surface morphology of the hydrogels by SEM. (**B**) Elemental composition and distribution of GelMA-PHEAA/CPC. (**C**) FT-IR spectra of GelMA, HEAA, GelMA-PHEAA, and GelMA-PHEAA/CPC. (**D**) XRD of CPC and bioactive hydrogel, and the characteristic peaks were labeled with *. (**E**) Typical compress-strain curves of different samples and detailed performance of the first 0–20% strain (**F**). (**G**) The Young's modulus of different hydrogels.

The stability of a hydrogel after swelling is very important because the mechanical property of traditional hydrogels is supposed to be weak in the swollen state. However, as shown in Figure 3A(I), the structure of GelMA-PHEAA/CPC hydrogel remained intact in the swollen state. We compressed the cylindrical hydrogel with a 500 g hook weight and it underwent deformation (Figure 3A(II,III)). After removing the force, the GelMA-PHEAA/CPC hydrogel recovered its initial shape, indicating that this hydrogel was stable in the solution environment. The swelling curve in Figure 3B exhibits the weight change of the freeze-dried hydrogels versus time. The swelling rate of pure GelMa hydrogel reached over 1000%, while, after introducing CPC or PHEAA in the system, the swelling rate of the hydrogels decreased. The PHEAA network is influenced more obviously than the inorganic component CPC. The GelMA-PHEAA/CPC hydrogel showed an appropriate ability to absorb water, and this behavior is thought to promote affinity with tissue. All hydrogels can be degraded by collagenase. The weight retention curves in Figure 3C demonstrated that the pure GelMA hydrogel degrades too fast to fill a defect, while the degradation ratio of GelMA/CPC was nearly 50% after 7 days. The hydrogel with PHEAA degraded more slowly than the single network materials, as the PHEAA promoted network density and enhanced the interaction of all components, making the system more stable. Thus, the GelMA-PHEAA/CPC needed a long time to degrade. This behavior could prolong the bioactive effect of the CPC hydrogel scaffold and slow the release of CPC, rather than rupturing quickly.

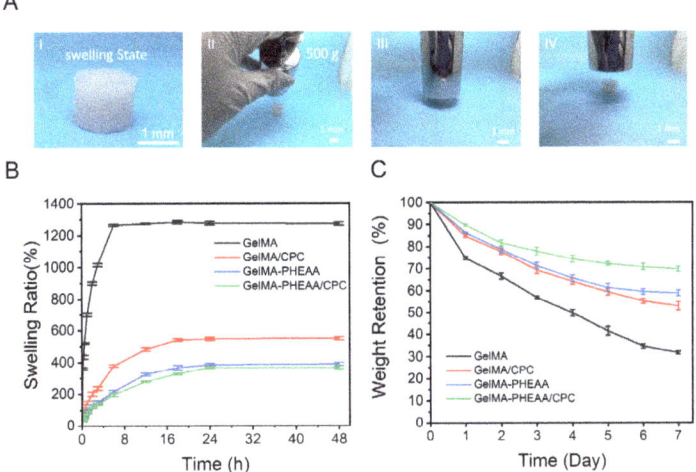

Figure 3. Stability test of the hydrogels. (**A**) The optical images of GelMA-PHEAA/CPC hydrogel in the swelling state. I: shape integrity of the swelling hydrogel. II–IV: the compression process on swollen GelMA-PHEAA/CPC hydrogels using a 500 g hook weight. (**B**) The swelling properties of different hydrogels. (**C**) Biodegradation of hydrogels.

2.2. Cell Proliferation

As presented in Figure 4, the Live/Dead staining showed almost no dead cells among all samples after being cultured for 48 and 72 h, suggesting that the biocompatibility of GelMA and the incorporation of CPC were satisfactory, and the hydrogels do not affect the proliferation of cells on the samples. L929 cells spread and grew well on all the sample surfaces cultured for 48 h. When cultured for 72 h, all of the samples were almost covered by cells. There were plenty of living cells (green fluorescence), indicating that all hydrogels were biocompatible, which is consistent with the previous literature [39] and demonstrates that all components are well-suited for clinical translation.

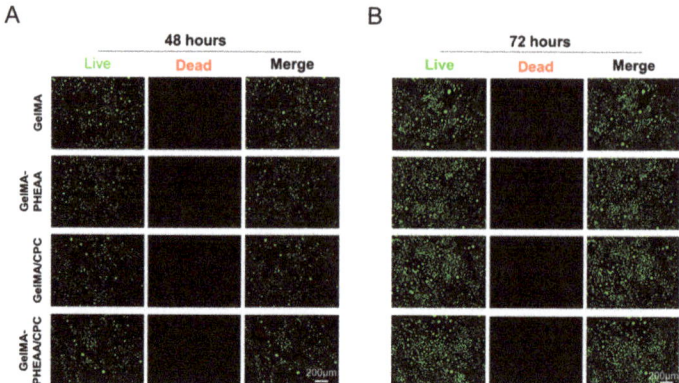

Figure 4. In vitro cell biocompatibility of GelMA-PHEAA/CPC hydrogels. Representative live/dead staining images of L929 cells for GelMA, GelMA-PHEAA, GelMA/CPC, and GelMA-PHEAA/CPC on 48 (**A**) and 72 h (**B**).

2.3. Osteogenic Activity

The expressions of osteogenic genes including Runx2, OPN, OCN, ALP, COL I, and OSX in MC3T3 were evaluated by qRT-PCR. As shown in Figure 5A–F, the cells on the GelMA-PHEAA/CPC hydrogel sample expressed a higher level of these osteogenic-related genes than the other samples. The trends in the expressions of the four genes in cells on the four samples were consistent and ran: GelMA < GelMA-PHEAA < GelMA/CPC < GelMA-PHEAA/CPC. The increased expressions of osteogenic genes in the cells may be due to the formation of a hard tissue structure and reduced risk of CPC rupture, and the good storage of CPC, which could promote the differentiation of MC3T3 cells.

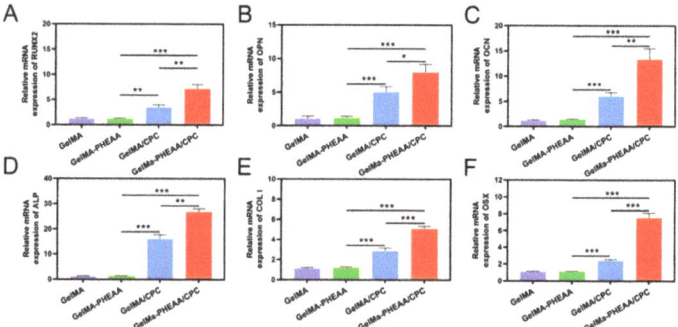

Figure 5. Expression of osteogenic-related genes in MC3T3 cells seeded on hydrogels at Day 7: (**A**) Runx2, (**B**) OPN, (**C**) OCN, (**D**) ALP, (**E**) COL I and (**F**) OSX, (n = 4, * $p < 0.05$, ** $p < 0.01$, *** $p < 0.001$).

Meanwhile, the protein expression of COL I and OCN, which were analyzed using immunofluorescence staining, were further investigated to evaluate the osteogenic activity of the hydrogels. As shown in Figure 6A,B, the GelMA-PHEAA/CPC hydrogel significantly promoted COL I and OCN expression compared to the other hydrogels. Semi-quantitative statistical analysis results further confirmed that the protein expression of COL I and OCN in the GelMA-PHEAA/CPC group was notably higher than that in the GelMA/CPC and GelMA hydrogels (Figure 5C,D). These results confirmed the bioactivity of the GelMA-PHEAA/CPC hydrogel. In summary, this bioactive injectable hydrogel is a novel promising therapy for fracture patients and well-suited to clinical commercialization.

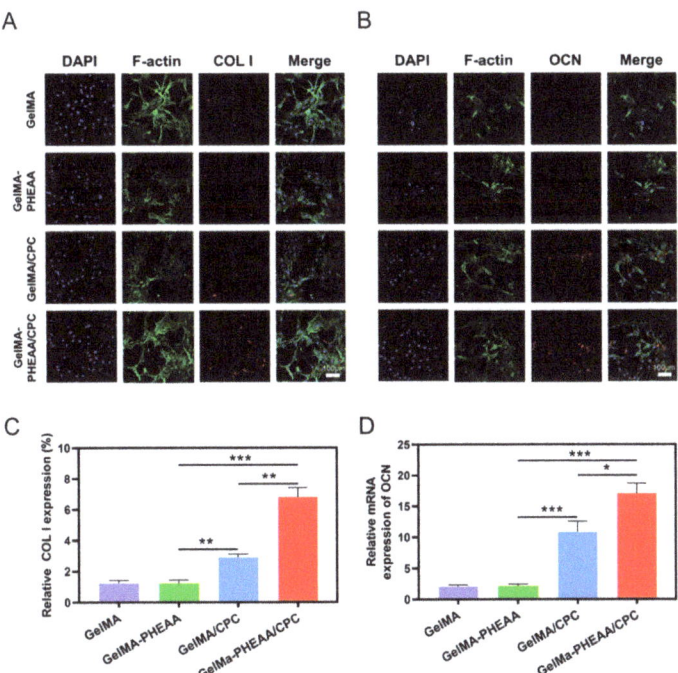

Figure 6. Immunofluorescence expressions stained by DAPI, F-actin, COL I (**A**), OCN (**B**), and their Merged images of GelMA, GelMA-PHEAA, GelMA/CPC, and GelMA-PHEAA/CPC on Days 21. Statistics of relative expression of COL I (**C**) and OCN (**D**) for GelMA, GelMA-PHEAA, GelMA/CPC, and GelMA-PHEAA/CPC in Days 21. The data was presented as the mean ± SD (n = 3), * $p < 0.05$, ** $p < 0.01$, *** $p < 0.001$.

3. Conclusions

In this work, GelMA-PHEAA/CPC hydrogel was synthesized through a one-pot process. Due to the fast cross-linking and injectable behavior, in situ defects can be formed easily. The GelMA-PHEAA network endowed the system with a strong and tough network, and the inorganic CPC phase promotes bioactivity for bone regeneration. The GelMA-PHEAA/CPC hydrogel exhibited good biocompatibility and promoted cell proliferation. The good storage of CPC in the hydrogel system promoted the mRNA expressions of osteogenic genes (Runx2, OPN, OCN, ALP, COL I, and OSX). The improved osteogenic activity of GelMA-PHEAA/CPC was due to the increase in the CPC content and the stable hydrogel system. This study provided a reference for the modulation synthesis of injectable hydrogel, strongly supporting the contention that the biological properties of CPC can be improved by modulation synthesis to endow bone implants with good osteogenic abilities.

4. Materials and Methods

4.1. Chemicals and Reagents

Methacrylate Gelatin (GelMA), N-Hydroxyethyl acrylamide (HEAA, 98%), and Lithium Phenyl(2,4,6-trimethylbenzoyl) phosphinate (LAP, 98%) were purchased from Aladdin Reagent Inc. (Shanghai). Tetracalcium phosphate (TTCP, $Ca_4(PO_4)_2O$) and dicalcium phosphate anhydrous (DCPA, $CaHPO_4$, 98%) were supplied by Macklin Biochemical Co., Ltd. (Shanghai, China). Calcium phosphate cement (CPC) powder was prepared by equimolar mixing of TTCP and DCPA. All reagents were used as received without further purification. In this experiment, all purified water was obtained from a Millipore system with an electronic conductance of 18.2 MΩ cm.

4.2. Preparation of GelMA-PHEAA/CPC Hydrogels

To obtain the fast gelation bioactive hydrogel, a predetermined amount of GelMA (0.3 g) was added to 1.5 mL purified water and stirred at 60 °C to prepare a GelMA solution. Then 1.2 g HEAA, 15 mg LAP photoinitiator, and 0.3 g CPC were added into the solution and subjected to ultrasonic dispersion (see Scheme 1). The obtained pre-solution was transferred into a syringe and injected into the template. After exposure to UV light (365 nm, 36 W) for 2 min, the bioactive hydrogel was fabricated and named GelMA-PHEAA/CPC. Hydrogels without CPC were defined as GelMA-PHEAA, and the 10 wt% GelMA hydrogel was chosen as a control group. The hydrogel base materials including GelMA, GelMA-PHEAA, GelMA/CPC, and GelMA-PHEAA/CPC were prepared by photo-initiation. Clinically used CPC scaffold was chosen as the control group [38].

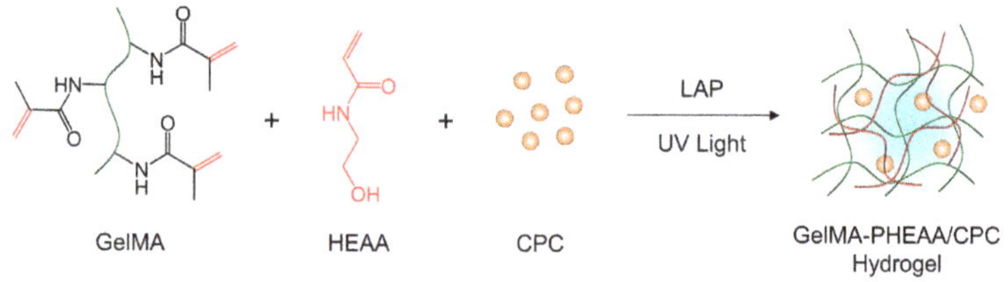

Scheme 1. Crosslinking process of GelMA-PHEAA/CPC hydrogel via UV polymerization.

4.3. Characterization of Hydrogels

Fourier-transform infrared (FTIR) spectra were acquired using a Nicolet 5700 (Thermo) at room temperature from 4000 to 400 cm^{-1}. The morphology and surface elemental composition of the hydrogels were visualized under scanning electron microscopy (SEM) (3400-N, Hitachi, Tokyo, Japan). The rheological behavior of the hydrogels was evaluated by a HAAKE MARS III rheometer. The pre-solution processability was tested under rotation ramp mode from 0.01–100 s^{-1} in 1 min at 37 °C. Dynamic frequency sweep tests were carried out from 15 to 0.1 Hz at 37 °C with an oscillatory strain of 1% at the thickness of 1 mm. The microstructure of the materials was examined by X-ray diffraction (XRD, Rigaku D/Max2550, Tokyo, Japan) with a scan range of 10 to 60 degrees. The mechanical properties of hydrogels were evaluated by an electronic mechanical testing machine (SANS CMT2503, Guangzhou, China). Hydrogel samples were fabricated in a cylindrical shape (8 mm in diameter and 10 mm in height) and tested at a speed of 10 mm min^{-1}. The swelling test was evaluated by gravimetric analysis. The freeze-dried hydrogel was weighed, giving W_d, and then hydrogels were immersed in phosphate-buffered saline (PBS). The hydrogels were taken out from PBS at different time intervals and weighed again, to find W_s, until swelling equilibrium. The swelling ratio was then calculated from swelling ratio = $(W_s - W_d)/W_d \times 100\%$. The degradation of the samples was also recorded using gravimetric analysis. The prepared hydrogels were weighed to find W_0 and then incubated in PBS with 2 CDU mL^{-1} collagenase type I solution at 37 °C for one week. The hydrogels were weighed every day to find W_t. The degradation ratio was then calculated from degradation ratio = $(W_0 - W_t)/W_0 \times 100\%$.

4.4. In Vitro Cytocompatibility Evaluation

All hydrogel specimens were immersed in sterile medium to reach a swelling equilibrium and further exposed under ultraviolet (UV) light (8 W) for another 1 h before testing. The hydrogel was soaked in fresh medium for 24 h to prepare the extracts. L929 cells were seeded in a 6-well plate for 24 h. Then, medium was replaced with extracts and incubated for 48 and 72 h. After culturing, the cells were stained with Calcein AM/PI (Servicebio, Beijing, China). Finally, the cells were viewed with fluorescence microscopy (Leica, Weztlar, Germany).

4.5. Quantitative Real-Time PCR (qRT-PCR) Analysis

The mRNA expressions of osteogenic genes in MC3T3-E1 cells on different samples were evaluated by using qRT-PCR. Briefly, MC3T3 cells with a cell density of 1×10^5 cells/mL were seeded on samples for 7 days. At the end of the incubation time, the cells were rinsed with PBS and the total RNA was extracted with TRIzol™ reagent (Invitrogen, MA, USA). Afterward, 1.0 µg of the RNA was reverse-transcribed into complementary DNA (cDNA) by Transcript or First Strand cDNA Synthesis Kit (Roche, Switzerland). Subsequently, qRT-PCR was carried out on the Roche LightCycler480 II system using an SYBR Green I PCR Master (Roche, Switzerland). The housekeeping gene was GAPDH, and runt-related transcription factor 2 (RUNX2), osteopontin (OPN), osteocalcin (OCN), Alkaline Phosphatase (ALP), type I collagen (COL I), and Osterix (OSX) were the chosen osteogenic genes. The relative mRNA expressions of target genes were normalized to that of the reference gene GAPDH. All the primers for RT-PCR are listed in Table 1.

Table 1. Primers for RT-PCR.

Primers		Sequences	Primers		Sequences
GAPDH	Forward	AGAACATCATCCCTGCATCCAC	GAPDH	Forward	
	Reverse	TCAGATCCACGACGGACACA		Reverse	
RUNX2	Forward	CCTCGAATGGCAGCACGCTA	RUNX2	Forward	
	Reverse	GCCGCCAAACAGACTCATCCA		Reverse	
ALP	Forward	CACGGCGTCCATGAGCAGAAC	ALP	Forward	
	Reverse	CAGGCACAGTGGTCAAGGTTGG		Reverse	
COL I	Forward	TGGTCCTGCTGGTCCTGCTG	COL I	Forward	
	Reverse	CTGTCACCTTGTTCGCCTGTCTC		Reverse	

4.6. Effects of the Osteogenic Activity of MC3T3

The osteogenic activity of the hydrogels was detected by immunofluorescence staining. Briefly, immunostaining of COL I and OCN was performed after 21 days of culture at a density of 1×10^4 MC3T3 cells per scaffold. After being fixed with 2.5% glutaraldehyde for 15 min, the cells were permeabilized with 0.1% Triton X-100 solution and blocked with 5% bovine serum albumin (BSA) for 1 h. Then, COL I and OCN were incubated with mouse-anti-osteocalcin IgG (Abcam, Cambridge, UK) at 4 °C overnight, followed by incubation with Alexa Fluor® 647 labeled goat-anti-mouse IgG (Abcam, HK, ab150115) for 2 h. Then, F-actin was stained with phalloidin, and the nucleus was stained with DAPI (Beyotime, Shanghai, China). Subsequently, the immunofluorescence images were observed and captured by a confocal laser scanning microscopy (CLSM, A1, Nikon, Natori, Japan).

4.7. Statistical Analysis

All numerical data were generated by at least three separate experiments and expressed as the mean and standard deviation of each experimental condition. One-way analysis of variance (ANOVA) was used in the statistical analysis, and Tukey's significant difference posterior test was used. Statistical significance was accepted at * $p < 0.05$, ** $p < 0.01$, and *** $p < 0.001$.

Author Contributions: Conceptualization, Y.W. and Z.P.; validation, Y.W. and Z.P.; writing—original draft preparation, D.Z., Y.W. and Z.P.; writing—review and editing, project administration, D.Z.; supervision, D.S. All authors have read and agreed to the published version of the manuscript.

Funding: This work is supported by the National Natural Science Foundation of China (No. 51972212, 81571828).

Data Availability Statement: The authors declare that all the data in the article are true and valid. If you need to quote from this article, please indicate the source.

Conflicts of Interest: The authors declare that there are no known competing financial interest or personal relationship that could have appeared to influence the work reported in this paper.

References

1. Williams, S.A.; Daigle, S.G.; Weiss, R.; Wang, Y.; Arora, T.; Curtis, J.R. Economic Burden of Osteoporosis-Related Fractures in the US Medicare Population. *Ann. Pharmacother.* **2021**, *55*, 821–829. [CrossRef] [PubMed]
2. Canalis, E.; Giustina, A.; Bilezikian, J.P. Mechanisms of anabolic therapies for osteoporosis. *N. Engl. J. Med.* **2007**, *357*, 905–916. [CrossRef] [PubMed]
3. Compston, J.E.; McClung, M.R.; Leslie, W.D. Osteoporosis. *Lancet* **2019**, *393*, 364–376. [CrossRef] [PubMed]
4. Johnell, O.; Kanis, J.A. An estimate of the worldwide prevalence and disability associated with osteoporotic fractures. *Osteoporos. Int.* **2006**, *17*, 1726–1733. [CrossRef]
5. van Geel, T.A.; van Helden, S.; Geusens, P.P.; Winkens, B.; Dinant, G.J. Clinical subsequent fractures cluster in time after first fractures. *Ann. Rheum. Dis.* **2009**, *68*, 99–102. [CrossRef]
6. Cummings, S.R.; Lui, L.-Y.; Eastell, R.; Allen, I.E. Association Between Drug Treatments for Patients with Osteoporosis and Overall Mortality Rates: A Meta-analysis. *JAMA Intern. Med.* **2019**, *179*, 1491–1500. [CrossRef]
7. Galbusera, F.; Volkheimer, D.; Reitmaier, S.; Berger-Roscher, N.; Kienle, A.; Wilke, H.-J. Pedicle screw loosening: A clinically relevant complication? *Eur. Spine J.* **2015**, *24*, 1005–1016. [CrossRef]
8. Weiser, L.; Huber, G.; Sellenschloh, K.; Viezens, L.; Püschel, K.; Morlock, M.M.; Lehmann, W. Insufficient stability of pedicle screws in osteoporotic vertebrae: Biomechanical correlation of bone mineral density and pedicle screw fixation strength. *Eur. Spine J.* **2017**, *26*, 2891–2897. [CrossRef]
9. Xu, Q.; Chen, Z.Y.; Zhang, Y.X.; Hu, X.F.; Chen, F.H.; Zhang, L.K.; Zhong, N.; Zhang, J.Y.; Wang, Y.B. Mussel-inspired bioactive 3D-printable poly(styrene-butadiene-styrene) and the in vitro assessment of its potential as cranioplasty implants. *J. Mater. Chem. B* **2022**, *10*, 3747–3758. [CrossRef]
10. Che, L.B.; Wang, Y.; Sha, D.Y.; Li, G.Y.; Wei, Z.H.; Liu, C.S.; Yuan, Y.; Song, D.W. A biomimetic and bioactive scaffold with intelligently pulsatile teriparatide delivery for local and systemic osteoporosis regeneration. *Bioact. Mater.* **2023**, *19*, 75–87. [CrossRef]
11. Che, L.B.; Lei, Z.Y.; Wu, P.Y.; Song, D.W. A 3D Printable and Bioactive Hydrogel Scaffold to Treat Traumatic Brain Injury. *Adv. Funct. Mater.* **2019**, *29*, 1904450. [CrossRef]
12. Wang, Q.; Dong, J.-F.; Fang, X.; Chen, Y. Application and modification of bone cement in vertebroplasty: A literature review. *Jt. Dis. Relat. Surg.* **2022**, *33*, 467–478. [CrossRef]
13. Saadeh, Y.S.; Swong, K.N.; Yee, T.J.; Strong, M.J.; Kashlan, O.N.; Szerlip, N.J.; Oppenlander, M.E.; Park, P. Effect of Fenestrated Pedicle Screws with Cement Augmentation in Osteoporotic Patients Undergoing Spinal Fusion. *World Neurosurg.* **2020**, *143*, e351–e361. [CrossRef]
14. Singh, V.; Mahajan, R.; Das, K.; Chhabra, H.S.; Rustagi, T. Surgical Trend Analysis for Use of Cement Augmented Pedicle Screws in Osteoporosis of Spine: A Systematic Review (2000–2017). *Glob. Spine J.* **2019**, *9*, 783–795. [CrossRef]
15. Raucci, M.G.; D'Amora, U.; Ronca, A.; Ambrosio, L. Injectable Functional Biomaterials for Minimally Invasive Surgery. *Adv. Healthc. Mater.* **2020**, *9*, 2000349. [CrossRef]
16. Zdorovets, M.V.; Kozlovskiy, A.L.; Borgekov, D.B.; Shlimas, D.I. Influence of irradiation with heavy Kr^{15+} ions on the structural, optical and strength properties of BeO ceramic. *J. Mater. Sci. Mater. Electron.* **2021**, *32*, 15375–15385. [CrossRef]
17. Trukhanov, S.V.; Trukhanov, A.V.; Turchenko, V.A.; Kostishyn, V.G.; Panina, L.V.; Kazakevich, I.S.; Balagurov, A.M. Structure and magnetic properties of $BaFe_{11.9}In_{0.1}O_{19}$ hexaferrite in a wide temperature range. *J. Alloys Compd.* **2016**, *689*, 383–393. [CrossRef]
18. Kozlovskiy, A.; Egizbek, K.; Darwish, M.V.; Ibragimova, M.; Shumskaya, A.; Rogachev, A.A.; Ignatovich, Z.V.; Kadyrzhanov, K. Evaluation of the Efficiency of Detection and Capture of Manganese in Aqueous Solutions of $FeCeO_x$ Nanocomposites Doped with Nb_2O_5. *Sensors* **2020**, *20*, 4851. [CrossRef]
19. Darwish, M.A.; Zubar, T.I.; Kanafyev, O.D.; Zhou, D.; Trukhanova, E.L.; Trukhanov, S.V.; Trukhanov, A.V.; Henaish, A.M. Combined Effect of Microstructure, Surface Energy, and Adhesion Force on the Friction of PVA/Ferrite Spinel Nanocomposites. *Nanomaterials* **2022**, *12*, 1998. [CrossRef]
20. Kozlovskiy, A.L.; Zdorovets, M.V. Effect of doping of $Ce^{4+/3+}$ on optical, strength and shielding properties of $(0.5-x)TeO_2$-$0.25MoO$-$0.25Bi_2O_3$-$xCeO_2$ glasses. *Mater. Chem. Phys.* **2021**, *263*, 124444. [CrossRef]
21. Almessiere, M.A.; Trukhanov, A.V.; Slimani, Y.; You, K.Y.; Trukhanov, S.V.; Trukhanova, E.L.; Esa, F.; Sadaqat, A.; Chaudhary, K.; Zdorovets, M.; et al. Correlation Between Composition and Electrodynamics Properties in Nanocomposites Based on Hard/Soft Ferrimagnetics with Strong Exchange Coupling. *Nanomaterials* **2019**, *9*, 202. [CrossRef] [PubMed]
22. Chen, F.; Song, Z.; Liu, C. Fast setting and anti-washout injectable calcium–magnesium phosphate cement for minimally invasive treatment of bone defects. *J. Mater. Chem. B* **2015**, *3*, 9173–9181. [CrossRef] [PubMed]
23. Ma, Y.; Zhang, W.; Wang, Z.; Wang, Z.; Xie, Q.; Niu, H.; Guo, H.; Yuan, Y.; Liu, C. PEGylated poly(glycerol sebacate)-modified calcium phosphate scaffolds with desirable mechanical behavior and enhanced osteogenic capacity. *Acta Biomater.* **2016**, *44*, 110–124. [CrossRef] [PubMed]
24. Wang, X.; Yu, Y.; Ji, L.; Geng, Z.; Wang, J.; Liu, C. Calcium phosphate-based materials regulate osteoclast-mediated osseointegration. *Bioact. Mater.* **2021**, *6*, 4517–4530. [CrossRef]
25. Cai, P.; Lu, S.; Yu, J.; Xiao, L.; Wang, J.; Liang, H.; Huang, L.; Han, G.; Bian, M.; Zhang, S.; et al. Injectable nanofiber-reinforced bone cement with controlled biodegradability for minimally-invasive bone regeneration. *Bioact. Mater.* **2023**, *21*, 267–283. [CrossRef]

26. Diaferia, C.; Rosa, E.; Balasco, N.; Sibillano, T.; Morelli, G.; Giannini, C.; Vitagliano, L.; Accardo, A. The Introduction of a Cysteine Residue Modulates The Mechanical Properties of Aromatic-Based Solid Aggregates and Self-Supporting Hydrogels. *Chem. Eur. J.* **2021**, *27*, 14886–14898. [CrossRef]
27. Rachmiel, D.; Anconina, I.; Rudnick-Glick, S.; Halperin-Sternfeld, M.; Adler-Abramovich, L.; Sitt, A. Hyaluronic Acid and a Short Peptide Improve the Performance of a PCL Electrospun Fibrous Scaffold Designed for Bone Tissue Engineering Applications. *Int. J. Mol. Sci.* **2021**, *22*, 2425. [CrossRef]
28. Wang, Y.; Ma, M.; Wang, J.; Zhang, W.; Lu, W.; Gao, Y.; Zhang, B.; Guo, Y. Development of a Photo-Crosslinking, Biodegradable GelMA/PEGDA Hydrogel for Guided Bone Regeneration Materials. *Materials* **2018**, *11*, 1345. [CrossRef]
29. Dong, Z.; Yuan, Q.; Huang, K.; Xu, W.; Liu, G.; Gu, Z. Gelatin methacryloyl (GelMA)-based biomaterials for bone regeneration. *RSC Adv.* **2019**, *9*, 17737–17744. [CrossRef]
30. Kurian, A.G.; Singh, R.K.; Patel, K.D.; Lee, J.-H.; Kim, H.-W. Multifunctional GelMA platforms with nanomaterials for advanced tissue therapeutics. *Bioact. Mater.* **2022**, *8*, 267–295. [CrossRef]
31. Pu, X.; Tong, L.; Wang, X.; Liu, Q.; Chen, M.; Li, X.; Lu, G.; Lan, W.; Li, Q.; Liang, J.; et al. Bioinspired Hydrogel Anchoring 3DP GelMA/HAp Scaffolds Accelerates Bone Reconstruction. *ACS Appl. Mater. Interfaces* **2022**, *14*, 20591–20602. [CrossRef]
32. Song, P.; Li, M.; Zhang, B.; Gui, X.; Han, Y.; Wang, L.; Zhou, W.; Guo, L.; Zhang, Z.; Li, Z.; et al. DLP fabricating of precision GelMA/HAp porous composite scaffold for bone tissue engineering application. *Compos. Part B Eng.* **2022**, *244*, 110163. [CrossRef]
33. Chen, H.; Liu, Y.; Ren, B.; Zhang, Y.; Ma, J.; Xu, L.; Chen, Q.; Zheng, J. Super Bulk and Interfacial Toughness of Physically Crosslinked Double-Network Hydrogels. *Adv. Funct. Mater.* **2017**, *27*, 1703086. [CrossRef]
34. Tang, L.; Zhang, D.; Gong, L.; Zhang, Y.; Xie, S.; Ren, B.; Liu, Y.; Yang, F.; Zhou, G.; Chang, Y.; et al. Double-Network Physical Cross-Linking Strategy To Promote Bulk Mechanical and Surface Adhesive Properties of Hydrogels. *Macromolecules* **2019**, *52*, 9512–9525. [CrossRef]
35. Zhang, D.; Tang, Y.; Zhang, Y.; Yang, F.; Liu, Y.; Wang, X.; Yang, J.; Gong, X.; Zheng, J. Highly stretchable, self-adhesive, biocompatible, conductive hydrogels as fully polymeric strain sensors. *J. Mater. Chem. A* **2020**, *8*, 20474–20485. [CrossRef]
36. Zhang, D.; Yang, F.; He, J.; Xu, L.; Wang, T.; Feng, Z.-Q.; Chang, Y.; Gong, X.; Zhang, G.; Zheng, J. Multiple Physical Bonds to Realize Highly Tough and Self-Adhesive Double-Network Hydrogels. *ACS Appl. Polym. Mater.* **2020**, *2*, 1031–1042. [CrossRef]
37. Sha, D.; Tang, S.; Dong, Z.; Chen, K.; Wang, N.; Liu, C.; Ling, X.; He, H.; Yuan, Y. Wearable, antibacterial, and self-healable modular sensors for monitoring joints movement ultra-sensitively. *Eur. Polym. J.* **2022**, *180*, 111617. [CrossRef]
38. Wu, F.; Wei, J.; Guo, H.; Chen, F.; Hong, H.; Liu, C. Self-setting bioactive calcium–magnesium phosphate cement with high strength and degradability for bone regeneration. *Acta Biomater.* **2008**, *4*, 1873–1884. [CrossRef]
39. Wang, S.G.; Wang, F.; Shi, K.; Yuan, J.F.; Sun, W.L.; Yang, J.T.; Chen, Y.X.; Zhang, D.; Che, L.B. Osteichthyes skin-inspired tough and sticky composite hydrogels for dynamic adhesive dressings. *Compos. Part B Eng.* **2022**, *241*, 110010. [CrossRef]

Disclaimer/Publisher's Note: The statements, opinions and data contained in all publications are solely those of the individual author(s) and contributor(s) and not of MDPI and/or the editor(s). MDPI and/or the editor(s) disclaim responsibility for any injury to people or property resulting from any ideas, methods, instructions or products referred to in the content.

Article

Cryogel System Based on Poly(vinyl alcohol)/Poly(ethylene brassylate-co-squaric acid) Platform with Dual Bioactive Activity

Bianca-Elena-Beatrice Crețu [1], Alina Gabriela Rusu [1], Alina Ghilan [1,*], Irina Rosca [2], Loredana Elena Nita [1] and Aurica P. Chiriac [1,*]

1. Department of Natural Polymers, Bioactive and Biocompatible Materials,
Petru Poni Institute of Macromolecular Chemistry, 41 A Grigore Ghica Voda Alley, 700487 Iasi, Romania
2. Center of Advanced Research in Bionanoconjugates and Biopolymers,
Petru Poni Institute of Macromolecular Chemistry, 41 A Grigore Ghica Voda Alley, 700487 Iasi, Romania
* Correspondence: diaconu.alina@icmpp.ro (A.G.); achiriac@icmpp.ro (A.P.C.)

Abstract: The inability to meet and ensure as many requirements as possible is fully justified by the continuous interest in obtaining new multifunctional materials. A new cryogel system based on poly(vinyl alcohol) (PVA) and poly(ethylene brassylate-co-squaric acid) (PEBSA) obtained by repeated freeze–thaw processes was previously reported and used for the incorporation of an antibacterial essential oil—namely, thymol (Thy). Furthermore, the present study aims to confer antioxidant properties to the PVA/PEBSA_Thy system by encapsulating α-tocopherol (α-Tcp), targeting a double therapeutic effect due to the presence of both bioactive compounds. The amphiphilic nature of the PEBSA copolymer allowed for the encapsulation of both Thy and α-Tcp, via an in situ entrapment method. The new PVA/PEBSA_Thy_α-Tcp systems were characterized in terms of their influence on the composition, network morphology and release profiles, as well as their antimicrobial and antioxidant properties. The study underlined the cumulative antioxidant efficiency of Thy and α-Tcp, which in combination with the PEBSA copolymer have a synergistic effect (97.1%). We believe that the convenient and simple strategy offered in this study increases applicability for these new PVA/PEBSA_Thy_α-Tcp cryogel systems.

Keywords: poly(vinyl alcohol); copolymacrolactone; thymol; α-tocopherol; synergistic effect; hydrogel dressings

Citation: Crețu, B.-E.-B.; Rusu, A.G.; Ghilan, A.; Rosca, I.; Nita, L.E.; Chiriac, A.P. Cryogel System Based on Poly(vinyl alcohol)/Poly(ethylene brassylate-co-squaric acid) Platform with Dual Bioactive Activity. *Gels* **2023**, *9*, 174. https://doi.org/10.3390/gels9030174

Academic Editors: Dong Zhang, Jintao Yang, Xiaoxia Le and Dianwen Song

Received: 6 February 2023
Revised: 20 February 2023
Accepted: 21 February 2023
Published: 22 February 2023

Copyright: © 2023 by the authors. Licensee MDPI, Basel, Switzerland. This article is an open access article distributed under the terms and conditions of the Creative Commons Attribution (CC BY) license (https://creativecommons.org/licenses/by/4.0/).

1. Introduction

A wound is defined as the disruption of the anatomical structure and normal function of the skin, caused by various types of trauma, burns, or surgery [1]. As the wound occurs, the healing process begins [2]. During this process, the generation of an excess of reactive oxygen species (ROS) also occurs as part of the defense mechanism against pathogens. ROSs are small molecules derived from unstable oxygen that try to stabilize themselves by capturing the electrons of some molecules in living organisms, implying the appearance of associated complications (dysfunctions at the level of cell membranes, conformational changes in proteins, loss of enzyme roles along with breaking DNA chains) [3,4].

Antioxidants are chemical compounds that can alleviate oxidative stress by donating electrons to other molecules, such as ROSs, and support the wound healing process [5]. Antioxidants can be classified as endogenous, produced naturally in the body as, for example, superoxide dismutase (SOD), catalase (CAT), glutathione peroxidase (GPx), and exogenous obtained through the diet, such as carotenoids, vitamin C, vitamin E and flavonoids, among others [6]. Among the tocopherols, α-Tcp is the most abundant form of vitamin E and is considered a powerful fat-soluble antioxidant that protects membrane lipids against oxidation and contributes to the mechanical stabilization of membranes through

Van der Waals physical interactions [7]. The administration of α-tocopherol, as well as the other liposoluble vitamins, presents some particularly important challenges due to its low water solubility and stability [8]. Moreover, vitamins are sensitive molecules so they require protection from pro-oxidant factors such as oxygen, UV or high temperatures. In this sense, the encapsulation of vitamins into a polymeric network could be a promising approach to preserve their chemical integrity and effectiveness, but also their controlled release, thus reducing the occurrence of hypervitaminosis syndrome [9]. Conventional hydrogels are commonly used by scientists to overcome the problems mentioned above, but the problems surrounding their implementation on a large scale increase due to the closed and small cavity of hydrogels. Cryogels have received tremendous attention in applications targeting the controlled release of active principles and tissue engineering of the skin, considering their large pore size, rough surface, absorption capacity and rapid swelling [10,11]. PVA is one of the most investigated water-soluble synthetic polymers in obtaining cryogels, with applications as drug carriers, in wound dressings and for tissue engineering due to its biocompatibility, biodegradability and non-toxicity [12].

A new cryogel system based on PVA and PEBSA obtained by repeated freeze–thaw processes was previously reported and used for the incorporation of an antibacterial essential oil—namely, Thy [13]. Thy, chemically known as 2-isopropyl-5-methylphenol, is a natural monoterpenoid phenol which is isolated from *Thymus vulgaris* and other plants such as *Ocimum gratissimum* L., *Origanum* L., *Satureja thymbra* L. [14]. Various pharmacological properties of thymol have been investigated and reported, including antimicrobial, antifungal, antioxidant, anti-inflammatory, analgesic, and healing activities [15–18]. PEBSA, a copolymacrolactone system was synthetized from ethylene brassylate (EB) and squaric acid (SA) by the ring-opening copolymerization procedure described before [19,20]. The supramolecular structure and high functionality of PEBSA copolymer, as well as its biocompatibility and good thermal stability, have led to it being recommended as a matrix for the incorporation of hydrophobic bioactive compounds [21,22].

In order to achieve a system with dual effect and activity, specifically antioxidant and antimicrobial, the alfa-tocopherol-thymol bioactive formulation was encapsulated in the new PVA/PEBSA polymer network. The PEBSA copolymer, due to its amphiphilic character, allowed for the encapsulation of hydrophobic bioactive substances such as Thy and α-Tcp, via an entrapment in situ method. Moreover, the hydrophobic affinity of the compounds involved in the system led to a good dispersion of the bioactive molecular agents in the PEBSA polymer network [20]. The newly prepared bioactive complexes were characterized in terms of the influence of the composition on the network morphology and release profiles, as well as the dual bioactive behavior due to the presence of Thy and α-Tcp. To support previous studies, PVA/PEBSA_Thy_α-Tcp systems were prepared using PEBSA with three different ratios between EB and SA comonomers, namely, 25/75, 50/50 and 75/25. To our knowledge, there have been no reported studies focusing on the synergistic effect between the antimicrobial properties of Thy and the antioxidant properties of α-Tcp associated with biomedical applications. Overall, this study offers a convenient strategy to achieve a PVA/PEBSA cryogel system with dual therapeutic effect due to the presence of both bioactive compounds. We intend to use the PVA/PEBSA_Thy_α-Tcp cryogel system for antimicrobial and wound healing applications, and in this regard, we carry out the necessary studies.

2. Results and Discussion

The present study aims to obtain cryogels based on PVA/PEBSA incorporating Thy and α-Tcp, targeting dual activity generated by the presence of the two bioactive compounds. Our group envisioned this approach to enhance the efficacy of these new antimicrobial cryogels to be used as wound dressings. The encapsulation of the bioactive compounds into the PEBSA polymeric matrix was realized by an inclusion complexation performed in 1,4-dioxane by entrapping of Thy, of α-Tcp, into the amphiphilic PEBSA network (Figure 1).

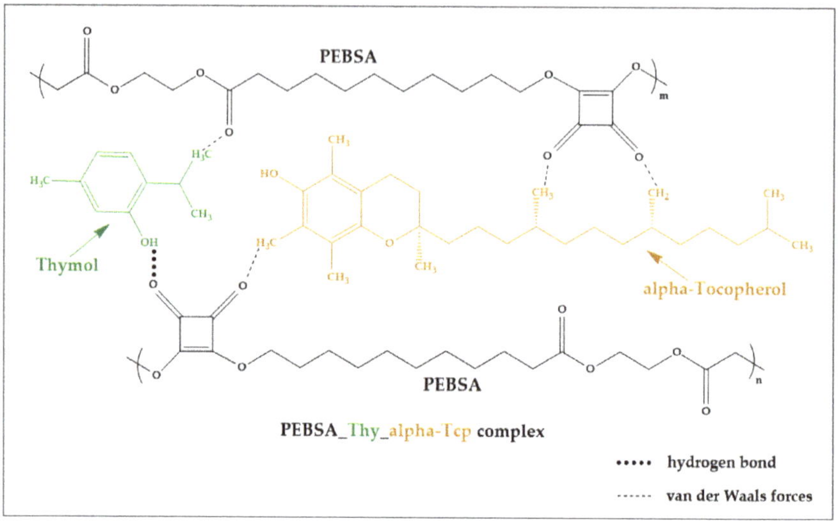

Figure 1. Illustration of PEBSA_Thy_α-Tcp bioactive complex formation.

2.1. Morphological Analysis

The internal morphology of the PVA/PEBSA polymer matrix and PVA/PEBSA_Thy_α-Tcp systems were evaluated by SEM. As can be seen in Figure 2A, the SEM micrographs of cryogels based on PVA and PEBSA copolymacrolactone illustrate their three-dimensional network with interconnected honeycomb-like pores and numerous meshes, which can ensure the incorporation of bioactive molecular compounds. The morphology of the PVA/PEBSA_Thy_α-Tcp systems highlights a relatively uniform distribution of the bioactive substances on the surface of the polymer network. The homogeneous distribution of Thy and α-Tcp in the PEBSA polymer network is due to the amphiphilic character of the copolymer and also to the hydrophobic affinity of the compounds involved in the system. Some differences in their morphology result from the cryogel preparation condition namely, the ratio between EB and SA comonomers in the copolymacrolactone system as well as the amount of bioactive compound.

Therefore, in the case of the PVA_PEBSA_Thy_α-Tcp sample (Figure 2B–D), with the increase in the EB/SA ratio in the sample, the pore size increases. This is associated with an increase in the number of carbonyl functional groups in SA, which further results in a higher number of hydrogen bonds between the mixing partners. Thus, the structure and pore size of PVA/PEBSA cryogels can be modified by adjusting the ratio between EB and SA comonomers from PEBSA composition. The morphological characterization of the new variants of cryogels is consistent with that recently reported in the literature [13].

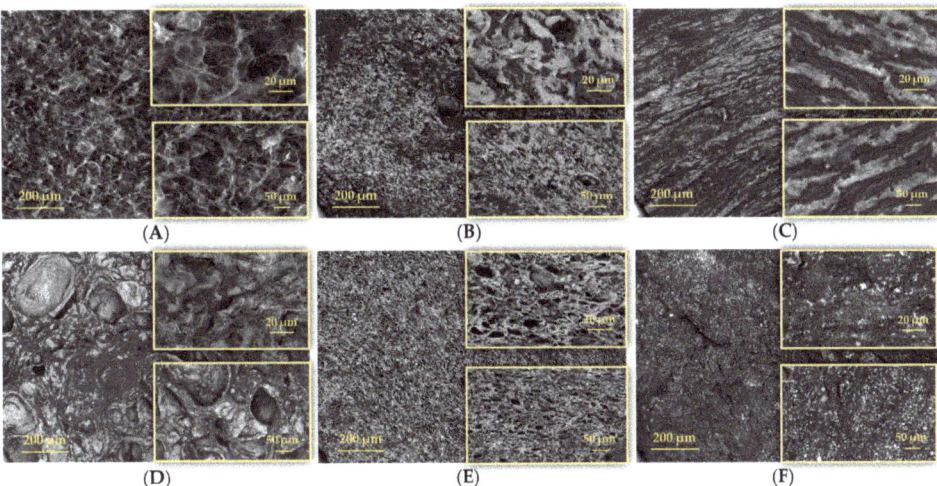

Figure 2. Comparative SEM images of the cryogels: (**A**) PVA_PEBSA$_{50/50}$, (**B**) PVA_PEBSA$_{25/75}$_Thy_α-Tcp, (**C**) PVA_PEBSA$_{50/50}$_Thy_α-Tcp, (**D**) PVA_PEBSA$_{75/25}$_Thy_α-Tcp, (**E**) PVA_PEBSA$_{50/50}$_2xThy_α-Tcp, and (**F**) PVA_PEBSA$_{50/50}$_Thy_2xα-Tcp.

2.2. Release Studies

The burst release of drugs from therapeutic formulations is not desirable, especially in the treatment of wound infections [23]. Therefore, incorporating these active principles into an effective delivery system, such as a hydrogel, and releasing them in a controlled manner reduces potential side effects. In the reported literature, there are studies that focused on the controlled release of Thy and α-Tcp from a hydrogel polymer matrix—examples are shown in Table 1:

Table 1. A summary of controlled release studies of bioactive compounds from hydrogels.

Scaffold Material	Bioactive Compounds		Cumulative Bioactive Compound Release in Different Release Media	Reference
	Thy	α-Tcp		
Chitosan hydrogels	✓	-	~ 70% of Thy release in artificial saliva and ~ 45% of Thy release in phosphate-buffered saline (PBS) after 4 h; 100% of Thy release in almost 48 h	[24]
Sodium alginate/chitosan hydrogels	-	✓	37.9 ± 5.18% of α-Tcp release in simulated body fluid after 24 h; a sustained release of 77.2 ± 11.51% over 14 days	[25]
Sodium alginate/poly(2-ethyl-2-oxazoline) chitosan-coated semi-interpenetrating hydrogels	✓	-	78.1 ± 1.7% of Thy release in PBS after 25 days	[26]
Hydroxypropyl-β-cyclodextrin hydrogels	✓	-	73.4–98.9% of Thy release in PBS, 7.4 after 6 h	[27]

Table 1. Cont.

Scaffold Material	Bioactive Compounds		Cumulative Bioactive Compound Release in Different Release Media	Reference
	Thy	α-Tcp		
Sodium caseinate/gelatin nanocomposite hydrogel	✓	-	71% of thyme essential oil in PBS, 7.4 supplemented with 20% ethanol after 72 h	[28]
PVA/pyrrolidone hydrogel	✓	-	70% of Thy release in ethanol solution after 5 days	[29]
Pluronic® F-127/nanocellulose hydrogel	-	✓	100% of α-Tcp release in 8 days	[30]

Starting from the premise that pH 5.4 is representative of healthy skin which can range between 5.4 and 8 when the deep layers of the skin are exposed following injuries [31], in this study, the cumulative release of the bioactive compounds was investigated at two different pHs: 5.4 and 7.4 (Figure 3).

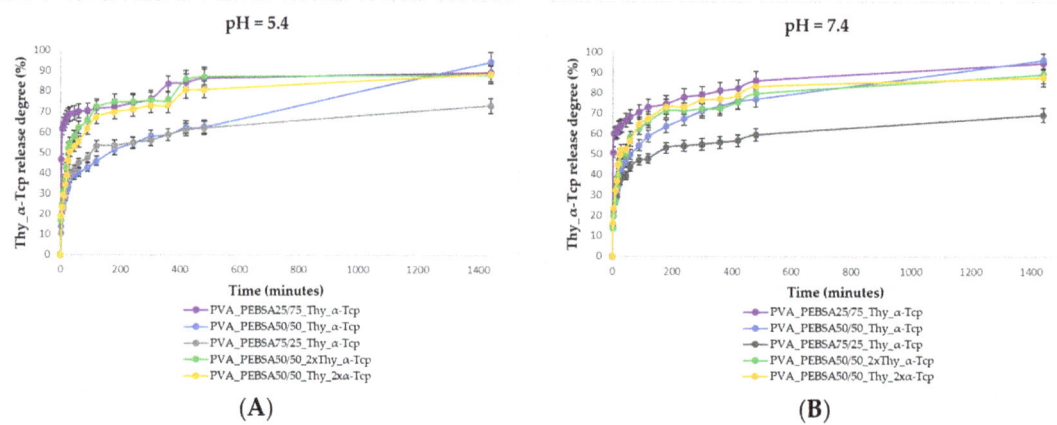

Figure 3. Cumulative release of Thy and α-Tcp at (A) pH 5.4 compared to (B) pH 7.4. Graphical data are expressed as mean ± standard error of the mean.

The initial burst release of Thy_α-Tcp complex from PVA_PEBSA$_{25/75}$_Thy_α-Tcp cryogel, 68.71% at pH 5.4 and 64% at pH 7.4, was observed in the first 30 min, while only 39.71% (pH 5.4) and 39.23% (pH 7.4) of bioactive compounds was released from PVA_PEBSA$_{75/25}$_Thy_α-Tcp cryogel at the same time. The release of a smaller amount of the bioactive substance during the burst release step in the case of the PVA_PEBSA$_{75/25}$_Thy_α-Tcp system correlates with a stronger hydrophobic character of PVA_PEBSA$_{75/25}$ matrix due to the higher ratio of EB. Therefore, the SA carbonyl groups ensure the coupling of bioactive compounds in the polymer matrix, while the hydrophobic alkyl chains of EB constitute the shell of the complex. Consequently, increasing the amount of EB comonomer to 75% in the chemical structure of the copolymer will determine the immobilization of the bioactive compounds in the network and their controlled release in a pulsating regime.

In order to study the influence of Thy_α-Tcp complex loading in the polymer matrix on the release capacity of the new bioactive compounds, the system with the equimolecular ratio between monomers was selected to vary the amount of bioactive compound (PEBSA$_{50/50}$_2xThy_α-Tcp and PEBSA$_{50/50}$_Thy_2xα-Tcp). The PVA_PEBSA$_{50/50}$_2xThy_α-Tcp and PVA_PEBSA$_{50/50}$_Thy_2xα-Tcp systems present a "burst" release of Thy and α-Tcp attributed to the diffusion of a double amount of bioactive compounds. According to Figure 3, the minimum and maximum release rates of Thy and α-Tcp from the studied systems over 24 h were 94.83% (pH 7.4) for the PVA_PEBSA$_{25/75}$_Thy_α-Tcp sample and 69.22% (pH 5.4) in the case of the PVA_PEBSA$_{75/25}$_Thy_α-Tcp sample. Since the largest number of granulocytes appear after 12–24 h at the injury site, the cells responsible for the immune response against microbial agents, the risk of infection is higher in the first minutes and hours after wounding. Consequently, the first 24 h after the appearance of the injury is the most important time interval to intervene with a material with antimicrobial properties to prevent infection. Therefore, the rate of drug release from hydrogel dressings is a significant factor in preventing infection [32].

2.3. Antimicrobial Activity

The antimicrobial activity screening of the newly synthesized bioactive compounds against *S. aureus* (Gram-positive bacterial strain), *C. albicans* (fungal strain), and *E. coli* (Gram-negative bacterial strain) was determined by disk diffusion assay. All the tested samples presented antimicrobial activity against the selected reference strains (as presented in Table 2 and Figures 4–6), results that correlate very well with recently reported antimicrobial assays [13].

Table 2. Antimicrobial activity (mm) of the tested samples against the reference strains.

Sample	Inhibition Zone (mm) *		
	S. aureus	*E. coli*	*C. albicans*
PVA_PEBSA$_{25/75}$_Thy_α-Tcp	22.30 ± 0.14	21.90 ± 0.99	38.55 ± 1.48
PVA_PEBSA$_{50/50}$_Thy_α-Tcp	21.10 ± 0.00	19.15 ± 1.06	32.25 ± 3.18
PVA_PEBSA$_{75/25}$_Thy_α-Tcp	25.90 ± 0.70	25.45 ± 3.46	34.65 ± 0.91
PVA_PEBSA$_{50/50}$_2xThy_α-Tcp	27.05 ± 0.63	28.40 ± 0.14	37.80 ± 0.28
PVA_PEBSA$_{50/50}$_Thy_2xα-Tcp	21.80 ± 3.25	19.20 ± 0.00	28.25 ± 0.77

* Data are represented as mean ± standard deviation from triplicate experiments.

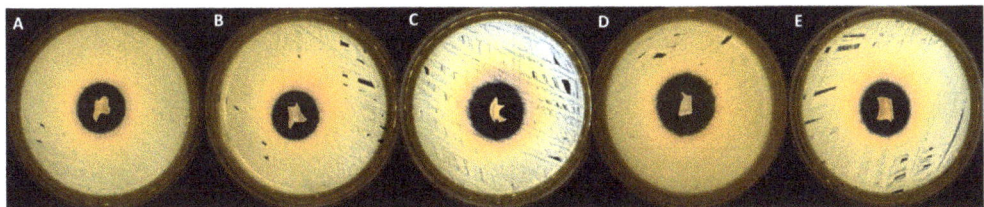

Figure 4. Antibacterial activity of the tested samples: (**A**) PVA_PEBSA$_{25/75}$_Thy_α-Tcp, (**B**) PVA_PEBSA$_{50/50}$_Thy_α-Tcp, (**C**) PVA_PEBSA$_{75/25}$_Thy_α-Tcp, (**D**) PVA_PEBSA$_{50/50}$_2xThy_α-Tcp, and (**E**) PVA_PEBSA$_{50/50}$_Thy_2xα-Tcp against *S. aureus*.

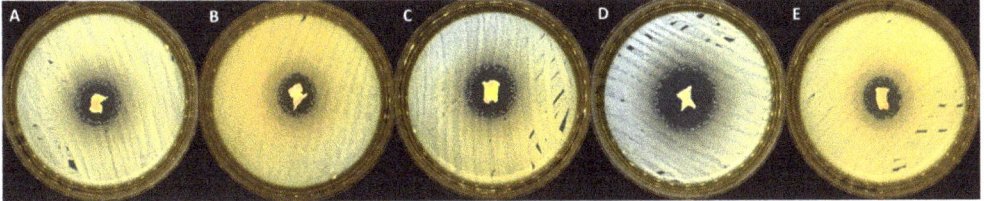

Figure 5. Antibacterial activity of the tested samples: (**A**) PVA_PEBSA$_{25/75}$_Thy_α-Tcp, (**B**) PVA_PEBSA$_{50/50}$_Thy_α-Tcp, (**C**) PVA_PEBSA$_{75/25}$_Thy_α-Tcp, (**D**) PVA_PEBSA$_{50/50}$_2xThy_α-Tcp, and (**E**) PVA_PEBSA$_{50/50}$_Thy_2xα-Tcp against *E. coli*.

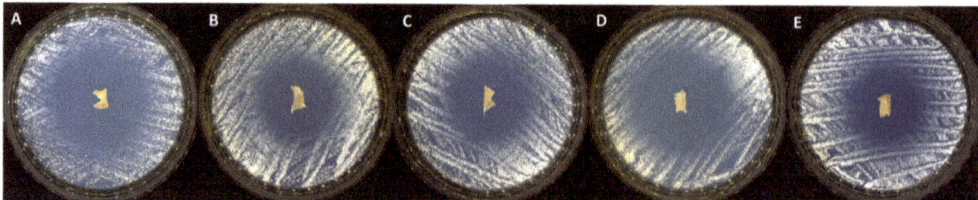

Figure 6. Antifungal activity of the tested samples: (**A**) PVA_PEBSA$_{25/75}$_Thy_α-Tcp, (**B**) PVA_PEBSA$_{50/50}$_Thy_α-Tcp, (**C**) PVA_PEBSA$_{75/25}$_Thy_α-Tcp, (**D**) PVA_PEBSA$_{50/50}$_2xThy_α-Tcp, and (**E**) PVA_PEBSA$_{50/50}$_Thy_2xα-Tcp against *C. albicans*.

The samples proved to be very effective especially against fungal strain represented by *C. albicans* (up to 38 mm of inhibition zone). Moreover, no significant differences were noticed in terms of antimicrobial activity among systems with different ratios between EB/SA comonomers. A smaller zone of inhibition was noticed in case of PVA_PEBSA$_{50/50}$_Thy_2xα-Tcp system against *E. coli* (19 mm), while the PVA_PEBSA$_{50/50}$_2xThy_α-Tcp system was more effective against all the tested microbial strains (>27 mm of inhibition zone)—Table 2, the efficiency related to the presence of Thy in a higher ratio. Compared to previous results [13], the addition of α-Tcp to the PVA$_{72000}$_PEBSA_Thy system did not substantially affect the antimicrobial character of the samples.

2.4. Antioxidant Efficiency

The interest of researchers in identifying new combinations of biomaterials that exhibit antimicrobial, antioxidant, anti-inflammatory, and healing activities for the treatment of wounds as well as their associated complications is increasing. α-Tcp, the most abundant form of vitamin E, is well known for its strong endogenous antioxidant activity by protecting membrane lipids against oxidation and mechanically stabilizing membranes, improving wound healing and skin regeneration [25]. Supplementing the PVA/PEBSA_Thy system with this antioxidant molecule targets a dual therapy and effect with these two bioactive compounds. The DPPH radical scavenging activity of the synthesized bioactive compounds is illustrated in Figure 7.

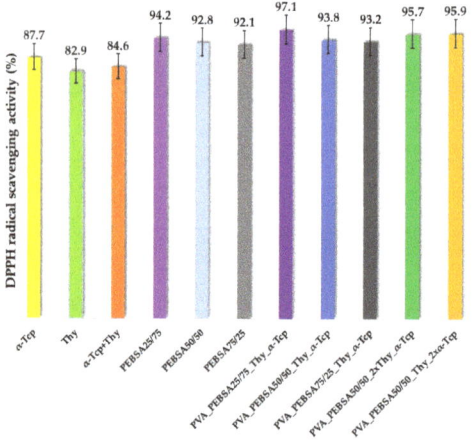

Figure 7. The antioxidant activity of the positive controls and PVA/PEBSA_Thy_α-Tcp systems. Graphical data are expressed as mean ± standard error of the mean.

According to the data obtained in this study, all samples showed antioxidant activity. The most evident activity is observed in the case of the PVA_PEBSA$_{25/75}$_Thy_α-Tcp system (97.1%) containing the PEBSA$_{25/75}$ variant. The carbonyl groups in SA have the ability to accept hydrogen; therefore, increasing the content of SA in the matrix determines a better free radical scavenging activity. At the same time, the cumulative antioxidant efficiency of Thy and α-Tcp in combination with the PEBSA copolymer has a remarkable synergistic effect, between 93.2% and 97.1% (Figure 7). In summary, the mixture of the two bioactive compounds and PEBSA produces new cryogels with potential applications as wound dressings whose therapeutic effects are superior to the effects produced by each individual component.

3. Conclusions

The encapsulation of hydrophobic molecular compounds into a polymer matrix has emerged as a method to modulate the low solubility in water and a promising approach to preserve their chemical integrity, efficacy, but also their controlled release in a pulsating or continuous regime. Three-dimensional scaffolds based on cryogels are strong candidates and of particular interest for this purpose [33]. In this study, our group used this strategy to develop new cryogels with antimicrobial and antioxidant activity based on PVA, PEBSA, Thy, and α-Tcp obtained by a repeated freeze–thaw process. On the one hand, Thy shows antimicrobial properties on a wide spectrum of bacteria (e.g., S. aureus, Bacillus licheniformis, E. coli, P. vulgaris, C. albicans, etc.) [16] and on the other hand, the encapsulation of α-Tcp confers antioxidant properties to the PVA/PEBSA_Thy system, targeting a double therapeutic effect due to the presence of both bioactive molecular agents. The new bioactive compounds prepared by encapsulation of Thy and α-Tcp into the PVA/PEBSA system were investigated from the point of view of the influence of the composition on the network morphology, release profiles, antimicrobial and antioxidant dual activity. SEM micrographs of the PVA/PEBSA polymer matrix illustrated their three-dimensional structure with interconnected pores and numerous meshes, which can ensure the incorporation of small molecular compounds. The morphology of PVA/PEBSA_Thy_α-Tcp systems highlights a relatively uniform distribution of the bioactive substances on the surface of the polymer network. The homogeneous distribution of Thy and α-Tcp into the PEBSA polymer network is due to the amphiphilic character of the copolymer and also to the hydrophobic affinity of the compounds involved in the system. Thy_α-Tcp release profiles from polymeric cryogels confirm the ability of PVA/PEBSA system to encapsulate these bioactive compounds. The lower release rate of Thy and α-Tcp during the burst release step in the case of the PVA_PEBSA$_{75/25}$_Thy_α-Tcp system (39.71% at pH 5.4 and 39.23% at pH 7.4) correlates with a stronger hydrophobic character of the PVA_PEBSA$_{75/25}$ matrix (due to a higher ratio of EB), determining the immobilization of the bioactive compounds in the network and their controlled release. The new PVA/PEBSA_Thy_α-Tcp systems proved antimicrobial activity against S. aureus (Gram-positive bacterial strain), C. albicans (fungal strain), and E. coli (Gram-negative bacterial strain). The study also underlined the cumulative antioxidant efficiency of Thy and α-Tcp, which in combination with the PEBSA copolymer have a synergistic effect (97.1%). However, this high potential of the investigated systems needs to be extensively evaluated by additional studies, in vitro and in vivo, with focus on possible cytotoxicity concerns, as well as their applicability in the management of skin wounds. The design of multifunctional hydrogel dressings with good adaptability to wounds and with painless on-demand removal property to avoid bacterial colonization remains a major problem to be solved [34,35].

4. Materials and Methods

4.1. Materials

Ethylene brassylate (EB, 1,4-dioxacycloheptadecane-5,17-dione, $C_{15}H_{26}O_4$, M_w = 270.36 g/mol, purity of 95.0%), squaric acid (SA, 3,4-dihydroxy-3-cyclobutene-1,2-dione, $H_2C_4O_4$, M_w = 114.06 g/mol, purity > 99.0%), (+)-α-Tocopherol (α-Tcp), 2,2-diphenyl-

1-picrylhydrazyl (DPPH), and 1,4-dioxane (purity ≥ 99.0%) were all purchased from Sigma-Aldrich (Darmstadt, Germany), poly(vinyl alcohol) (PVA, M_w = 72,000 g/mol, 98% hydrolyzed) was acquired from Merck (Hohenbrunn, Germany), thymol (Thy, 2-isopropyl-5-methylphenol, $C_{10}H_{14}O$) was obtained from Alfa Aesar (Kandel, Germany), anhydrous 1-hexanol was purchased from Across-Organics (Geel, Belgium), monosodium phosphate ($NaH_2PO_4 \times 2H_2O$). Disodium phosphate ($Na_2H_2PO_4 \times 7H_2O$) was procured from Chemical Company (Iasi, Romania) and ethanol (absolute, ≥99.8%) from Honeywell (Seelze, Germany). All chemicals were used as received without further purification.

4.2. Preparation of Cryogels by In Situ Entrapment of Thymol and α-Tocopherol

The cryogels were individually obtained by mixing proper ratios of PVA and PEBSA_Thy_α-Tcp complex solutions (Table 3), which were poured into molds and then subjected to three consecutive freeze–thaw cycles, respectively, freezing for 18 h at −20 °C followed by thawing for 8 h at 25 °C (ambient temperature) [13].

Table 3. Samples name and bioactive compound preparation.

Sample	PVA/PEBSA Ratio	Composition for a Volume of 5 mL Sample			
		PVA (g)	PEBSA (g)	Thymol (g)	α-Tocopherol (g)
PVA_PEBSA$_{25/75}$_Thy_α-Tcp	2/1	0.132	0.066	0.066	0.066
PVA_PEBSA$_{50/50}$_Thy_α-Tcp				0.066	0.066
PVA_PEBSA$_{75/25}$_Thy_α-Tcp				0.066	0.066
PVA_PEBSA$_{50/50}$_2xThy_α-Tcp				0.132	0.066
PVA_PEBSA$_{50/50}$_Thy_2xα-Tcp				0.066	0.132

Briefly, PEBSA was synthesized as previously described [19] by a polycondensation procedure of EB macrolactone after ring-opening with SA. The new bioactive compound was obtained by the initial preparation of the PEBSA_Thy_α-Tcp complex produced by mixing PEBSA (0.066 g/mL in 1,4-dioxane) with different amounts of Thy and α-Tcp to obtain the desired PEBSA/Thy/α-Tcp mass ratio (either 1/1/1, 1/2/1, or 1/1/2 $w/w/w$). Then, the obtained complex was mixed with the PVA solution (4% w/v) in a volumetric ratio of 2/1. The synthesized samples were frozen with liquid nitrogen and lyophilized for 24 h at −55 °C (Alpha 1-2LD Plus, Martin Christ, Germany) for further characterization.

4.3. Characterization

4.3.1. Morphological Analysis

The morphology in the cross-sections of the freeze-dried samples was observed by scanning electron microscopy (SEM Quanta 200, FEI Company, Hillsboro, OR, USA). The instrument operated with secondary electrons at 20 kV in low-vacuum mode, without any coating. Before analysis, the samples were fixed on aluminum stubs with double-adhesive carbon tape.

4.3.2. Release Studies

To study the in vitro release behavior of the bioactive compounds, each sample was weighed (20 mg) and incubated in 10 mL of PBS, 0.01 M, at a constant temperature of 37 °C for 24 h. The release profiles of the bioactive substances were measured under different conditions using buffer solutions of pH 5.4 and 7.4 to simulate the pH of normal healthy skin and, respectively, the pH of injured skin. At predetermined time intervals, 2 mL of each sample was extracted and analyzed at 283 nm using a UV-VIS spectrophotometer (Jenway 6305, Stone, Staffordshire, United Kingdom). The cumulative release of Thy and α-Tcp was calculated based on the calibration curves determined at the same wavelengths.

4.3.3. Antimicrobial Activity

The antimicrobial activity of the PVA/PEBSA_Thy_α-Tcp systems was determined using a disk diffusion assay [36,37] against three different reference strains: Gram-positive bacterial strain, *Staphylococcus aureus* ATCC25923 (*S. aureus*); Gram-negative bacterial strain, *Escherichia coli* ATCC25922 (*E. coli*); and fungal strain, *Candida albicans* ATCC10231 (*C. albicans*). All microorganisms were stored at −80 °C in 20% glycerol. The bacterial strains were refreshed on trypticase soy agar (TSA) at 37 °C and the yeast strain was refreshed on Sabouraud dextrose agar (SDA) at 37 °C. Microbial suspensions were prepared with these cultures in sterile solution to obtain turbidity optically comparable to that of 0.5 McFarland standards. Volumes of 0.1 mL from each inoculum were spread onto TSA/SDA plates, and then the sterilized samples of 10 mm and 25 mg each were added.

To evaluate the antimicrobial properties, the growth inhibition was measured under standard conditions after 24 h of incubation at 37 °C. All tests were carried out in triplicate for each sample. After incubation, the samples were analyzed with SCAN1200®, version 8.6.10.0 (Interscience, Saint Nom la Brétèche - FRANCE).

4.3.4. Antioxidant Efficiency

The free radical scavenging activity of the bioactive compounds was evaluated by the 2,2-diphenyl-1-picrylhydrazyl (DPPH) assay according to the methodology described by Brand-Williams et al. [38]. Briefly, 3 mL of ethanol and 20 mg of the sample were added to 0.3 mL of DPPH stock solution 0.5 mM in absolute ethanol. The control solution was prepared by mixing ethanol (3 mL) and DPPH stock solution (0.3 mL). The reaction mixture was incubated in a dark place at room temperature for 30 min. The changes in color, from intense violet to light yellow, were recorded spectrophotometrically at 517 nm (Jenway 6305 UV–VIS Spectrophotometer, Stone, Staffordshire, UK). The percentage of DPPH radical scavenging activity was calculated by the following equation:

$$\% \text{ DPPH radical scavenging activity} = \frac{A_C - A_S}{A_C} \times 100$$

where A_c is the absorbance of the control DPPH solution and A_S is the absorbance of the DPPH solution containing samples; the values reported for each sample represents the mean of three independent measurements.

4.3.5. Statistical Analysis

All experimental data were performed in triplicate and the results were expressed as mean ± standard error of the mean. Statistical analysis was performed with XLSTAT Ecology version 2019.4.1 software [39].

Author Contributions: Conceptualization, B.-E.-B.C. and L.E.N.; methodology, B.-E.-B.C., A.G.; validation, A.P.C., L.E.N., A.G.R.; formal analysis, B.-E.-B.C., I.R.; investigation, B.-E.-B.C.; resources, A.G.R.; data curation; writing—original draft preparation, B.-E.-B.C.; writing—review and editing, L.E.N.; visualization A.G.; supervision, A.P.C. and L.E.N.; project administration, A.G.R.; funding acquisition, A.G.R. All authors have read and agreed to the published version of the manuscript.

Funding: This work was financially supported by the grant of the Romanian National Authority for Scientific Research, CNCS-UEFISCDI, project number 657PED from 21/06/2022, PN-III-P2-2.1-PED-2021-2229 "Hybrid bio-systems enriched with biotechnological extracted oils and applicability in skin tissue engineering", within PNCDI III.

Conflicts of Interest: The authors declare no conflict of interest.

References

1. Enoch, S.; Leaper, D.J. Basic Science of Wound Healing. *Surgery* **2008**, *26*, 31–37. [CrossRef]
2. Velnar, T.; Bailey, T.; Smrkolj, V. The Wound Healing Process: An Overview of the Cellular and Molecular Mechanisms. *J. Int. Med. Res.* **2009**, *37*, 1528–1542. [CrossRef] [PubMed]

3. Aguilar, T.A.F.; HernándezNavarro, B.C.; Pérez, J.A.M.; Aguilar, T.A.F.; HernándezNavarro, B.C.; Pérez, J.A.M. *Endogenous Antioxidants: A Review of Their Role in Oxidative Stress*; IntechOpen: London, UK, 2016; ISBN 978-953-51-2838-0.
4. Na, Y.; Woo, J.; Choi, W.I.; Lee, J.H.; Hong, J.; Sung, D. α-Tocopherol-Loaded Reactive Oxygen Species-Scavenging Ferrocene Nanocapsules with High Antioxidant Efficacy for Wound Healing. *Int. J. Pharm.* **2021**, *596*, 120205. [CrossRef]
5. Comino-Sanz, I.M.; López-Franco, M.D.; Castro, B.; Pancorbo-Hidalgo, P.L. The Role of Antioxidants on Wound Healing: A Review of the Current Evidence. *JCM* **2021**, *10*, 3558. [CrossRef] [PubMed]
6. Roehrs, M.; Valentini, J.; Paniz, C.; Moro, A.; Charão, M.; Bulcão, R.; Freitas, F.; Brucker, N.; Duarte, M.; Leal, M.; et al. The Relationships between Exogenous and Endogenous Antioxidants with the Lipid Profile and Oxidative Damage in Hemodialysis Patients. *BMC Nephrol.* **2011**, *12*, 59. [CrossRef] [PubMed]
7. Srivastava, S.; Phadke, R.S.; Govil, G.; Rao, C.N.R. Fluidity, Permeability and Antioxidant Behaviour of Model Membranes Incorporated with α-Tocopherol and Vitamin E Acetate. *Biochim. Et Biophys. Acta (BBA)-Biomembr.* **1983**, *734*, 353–362. [CrossRef]
8. Bonferoni, M.C.; Riva, F.; Invernizzi, A.; Dellera, E.; Sandri, G.; Rossi, S.; Marrubini, G.; Bruni, G.; Vigani, B.; Caramella, C.; et al. Alpha Tocopherol Loaded Chitosan Oleate Nanoemulsions for Wound Healing. Evaluation on Cell Lines and Ex Vivo Human Biopsies, and Stabilization in Spray Dried Trojan Microparticles. *Eur. J. Pharm. Biopharm.* **2018**, *123*, 31–41. [CrossRef]
9. Gonnet, M.; Lethuaut, L.; Boury, F. New Trends in Encapsulation of Liposoluble Vitamins. *J. Control. Release* **2010**, *146*, 276–290. [CrossRef]
10. Haleem, A.; Chen, S.-Q.; Ullah, M.; Siddiq, M.; He, W.-D. Highly Porous Cryogels Loaded with Bimetallic Nanoparticles as an Efficient Antimicrobial Agent and Catalyst for Rapid Reduction of Water-Soluble Organic Contaminants. *J. Environ. Chem. Eng.* **2021**, *9*, 106510. [CrossRef]
11. Ambreen, J.; Haleem, A.; Shah, A.A.; Mushtaq, F.; Siddiq, M.; Bhatti, M.A.; Shah Bukhari, S.N.U.; Chandio, A.D.; Mahdi, W.A.; Alshehri, S. Facile Synthesis and Fabrication of NIPAM-Based Cryogels for Environmental Remediation. *Gels* **2023**, *9*, 64. [CrossRef]
12. Wang, M.; Bai, J.; Shao, K.; Tang, W.; Zhao, X.; Lin, D.; Huang, S.; Chen, C.; Ding, Z.; Ye, J. Poly(Vinyl Alcohol) Hydrogels: The Old and New Functional Materials. *Int. J. Polym. Sci.* **2021**, *2021*, 2225426. [CrossRef]
13. Nita, L.E.; Crețu, B.-E.-B.; Șerban, A.-M.; Rusu, A.G.; Rosca, I.; Pamfil, D.; Chiriac, A.P. New Cryogels Based on Poly (Vinyl Alcohol) and a Copolymacrolactone System. II. Antibacterial Properties of the Network Embedded with Thymol Bioactive Agent. *React. Funct. Polym.* **2023**, *182*, 105461. [CrossRef]
14. Nagoor Meeran, M.F.; Javed, H.; Al Taee, H.; Azimullah, S.; Ojha, S.K. Pharmacological Properties and Molecular Mechanisms of Thymol: Prospects for Its Therapeutic Potential and Pharmaceutical Development. *Front. Pharmacol.* **2017**, *8*, 380. [CrossRef] [PubMed]
15. Braga, P.C.; Dal Sasso, M.; Culici, M.; Bianchi, T.; Bordoni, L.; Marabini, L. Anti-Inflammatory Activity of Thymol: Inhibitory Effect on the Release of Human Neutrophil Elastase. *Pharmacology* **2006**, *77*, 130–136. [CrossRef]
16. Marchese, A.; Orhan, I.E.; Daglia, M.; Barbieri, R.; Di Lorenzo, A.; Nabavi, S.F.; Gortzi, O.; Izadi, M.; Nabavi, S.M. Antibacterial and Antifungal Activities of Thymol: A Brief Review of the Literature. *Food Chem.* **2016**, *210*, 402–414. [CrossRef]
17. Escobar, A.; Pérez, M.; Romanelli, G.; Blustein, G. Thymol Bioactivity: A Review Focusing on Practical Applications. *Arab. J. Chem.* **2020**, *13*, 9243–9269. [CrossRef]
18. Najafloo, R.; Behyari, M.; Imani, R.; Nour, S. A Mini-Review of Thymol Incorporated Materials: Applications in Antibacterial Wound Dressing. *J. Drug Deliv. Sci. Technol.* **2020**, *60*, 101904. [CrossRef]
19. Chiriac, A.P.; Nita, L.E.; Macsim, A.-M.; Tudorachi, N.; Rosca, I.; Stoica, I.; Tampu, D.; Aflori, M.; Doroftei, F. Synthesis of Poly(Ethylene Brassylate-Co-Squaric Acid) as Potential Essential Oil Carrier. *Pharmaceutics* **2021**, *13*, 477. [CrossRef]
20. Crețu, B.-E.-B.; Nita, L.E.; Șerban, A.-M.; Rusu, A.G.; Doroftei, F.; Chiriac, A.P. New Cryogels Based on Poly(Vinyl Alcohol) and a Copolymacrolactone System: I-Synthesis and Characterization. *Nanomaterials* **2022**, *12*, 2420. [CrossRef] [PubMed]
21. Chiriac, A.P.; Asandulesa, M.; Stoica, I.; Tudorachi, N.; Rusu, A.G.; Nita, L.E.; Chiriac, V.M.; Timpu, D. Comparative Study on the Properties of a Bio-Based Copolymacrolactone System. *Polym. Test.* **2022**, *109*, 107555. [CrossRef]
22. Chiriac, A.P.; Stoleru, E.; Rosca, I.; Serban, A.; Nita, L.E.; Rusu, A.G.; Ghilan, A.; Macsim, A.-M.; Mititelu-Tartau, L. Development of a New Polymer Network System Carrier of Essential Oils. *Biomed. Pharmacother.* **2022**, *149*, 112919. [CrossRef]
23. Rusu, A.G.; Chiriac, A.P.; Nita, L.E.; Ghilan, A.; Rusu, D.; Simionescu, N.; Tartau, L.M. Nanostructured Hyaluronic Acid-Based Hydrogels Encapsulating Synthetic/ Natural Hybrid Nanogels as Promising Wound Dressings. *Biochem. Eng. J.* **2022**, *179*, 108341. [CrossRef]
24. Alvarez Echazú, M.I.; Olivetti, C.E.; Anesini, C.; Perez, C.J.; Alvarez, G.S.; Desimone, M.F. Development and Evaluation of Thymol-Chitosan Hydrogels with Antimicrobial-Antioxidant Activity for Oral Local Delivery. *Mater. Sci. Eng. C* **2017**, *81*, 588–596. [CrossRef] [PubMed]
25. Ehterami, A.; Salehi, M.; Farzamfar, S.; Samadian, H.; Vaez, A.; Ghorbani, S.; Ai, J.; Sahrapeyma, H. Chitosan/Alginate Hydrogels Containing Alpha-Tocopherol for Wound Healing in Rat Model. *J. Drug Deliv. Sci. Technol.* **2019**, *51*, 204–213. [CrossRef]
26. Lavanya, K.; Balagangadharan, K.; Chandran, S.V.; Selvamurugan, N. Chitosan-Coated and Thymol-Loaded Polymeric Semi-Interpenetrating Hydrogels: An Effective Platform for Bioactive Molecule Delivery and Bone Regeneration in Vivo. *Biomater. Adv.* **2023**, *146*, 213305. [CrossRef]

27. Garg, A.; Ahmad, J.; Hassan, M.Z. Inclusion Complex of Thymol and Hydroxypropyl-β-Cyclodextrin (HP-β-CD) in Polymeric Hydrogel for Topical Application: Physicochemical Characterization, Molecular Docking, and Stability Evaluation. *J. Drug Deliv. Sci. Technol.* **2021**, *64*, 102609. [CrossRef]
28. Alsakhawy, S.A.; Baghdadi, H.H.; El-Shenawy, M.A.; Sabra, S.A.; El-Hosseiny, L.S. Encapsulation of Thymus Vulgaris Essential Oil in Caseinate/Gelatin Nanocomposite Hydrogel: In Vitro Antibacterial Activity and in Vivo Wound Healing Potential. *Int. J. Pharm.* **2022**, *628*, 122280. [CrossRef]
29. Malka, E.; Caspi, A.; Cohen, R.; Margel, S. Fabrication and Characterization of Hydrogen Peroxide and Thymol-Loaded PVA/PVP Hydrogel Coatings as a Novel Anti-Mold Surface for Hay Protection. *Polymers* **2022**, *14*, 5518. [CrossRef]
30. Afrin Shefa, A.; Park, M.; Gwon, J.-G.; Lee, B.-T. Alpha Tocopherol-Nanocellulose Loaded Alginate Membranes and Pluronic Hydrogels for Diabetic Wound Healing. *Mater. Des.* **2022**, *224*, 111404. [CrossRef]
31. Jones, E.M.; Cochrane, C.A.; Percival, S.L. The Effect of PH on the Extracellular Matrix and Biofilms. *Adv. Wound Care* **2015**, *4*, 431–439. [CrossRef]
32. Darabian, B.; Bagheri, H.; Mohammadi, S. Improvement in Mechanical Properties and Biodegradability of PLA Using Poly(Ethylene Glycol) and Triacetin for Antibacterial Wound Dressing Applications. *Prog. Biomater.* **2020**, *9*, 45–64. [CrossRef]
33. You, S.; Huang, Y.; Mao, R.; Xiang, Y.; Cai, E.; Chen, Y.; Shen, J.; Dong, W.; Qi, X. Together Is Better: Poly(Tannic Acid) Nanorods Functionalized Polysaccharide Hydrogels for Diabetic Wound Healing. *Ind. Crops Prod.* **2022**, *186*, 115273. [CrossRef]
34. Li, Y.; Fu, R.; Duan, Z.; Zhu, C.; Fan, D. Adaptive Hydrogels Based on Nanozyme with Dual-Enhanced Triple Enzyme-Like Activities for Wound Disinfection and Mimicking Antioxidant Defense System. *Adv. Healthc. Mater.* **2022**, *11*, 2101849. [CrossRef]
35. Yang, Y.; Xu, H.; Li, M.; Li, Z.; Zhang, H.; Guo, B.; Zhang, J. Antibacterial Conductive UV-Blocking Adhesion Hydrogel Dressing with Mild On-Demand Removability Accelerated Drug-Resistant Bacteria-Infected Wound Healing. *ACS Appl. Mater. Interfaces* **2022**, *14*, 41726–41741. [CrossRef] [PubMed]
36. Bauer, A.W.; Perry, D.M.; Kirby, W.M. Single-Disk Antibiotic-Sensitivity Testing of Staphylococci; an Analysis of Technique and Results. *AMA Arch. Intern. Med.* **1959**, *104*, 208–216. [CrossRef] [PubMed]
37. Clinical and Laboratory Standards Institute (CLSI). *Performance Standards for Anti-Microbial Susceptibility Testing*, 32nd ed.; CLSI Supplement M100 (ISBN 978-1-68440-134-5 [Print]; ISBN 978-1-68440-135-2 [Electronic]); Clinical and LaboraTory Standards Institute: Malvern, PA, USA, 2022.
38. Brand-Williams, W.; Cuvelier, M.E.; Berset, C. Use of a Free Radical Method to Evaluate Antioxidant Activity. *LWT-Food Sci. Technol.* **1995**, *28*, 25–30. [CrossRef]
39. XLSTAT. Statistical Software for Excel. Available online: https://www.xlstat.com/en/ (accessed on 28 November 2022).

Disclaimer/Publisher's Note: The statements, opinions and data contained in all publications are solely those of the individual author(s) and contributor(s) and not of MDPI and/or the editor(s). MDPI and/or the editor(s) disclaim responsibility for any injury to people or property resulting from any ideas, methods, instructions or products referred to in the content.

Article

Thermosensitive Shape-Memory Poly(stearyl acrylate-*co*-methoxy poly(ethylene glycol) acrylate) Hydrogels

Hideaki Tokuyama *, Ryo Iriki and Makino Kubota

Department of Chemical Engineering, Tokyo University of Agriculture and Technology, Tokyo 184-8588, Japan
* Correspondence: htoku@cc.tuat.ac.jp; Tel.: +81-42-388-7607

Abstract: Stimuli-sensitive hydrogels are highly desirable candidates for application in intelligent biomaterials. Thus, a novel thermosensitive hydrogel with shape-memory function was developed. Hydrophobic stearyl acrylate (SA), hydrophilic methoxy poly(ethylene glycol) acrylate (MPGA), and a crosslinking monomer were copolymerized to prepare poly(SA-*co*-MPGA) gels with various mole fractions of SA (x_{SA}) in ethanol. Subsequently, the prepared gels were washed, dried, and re-swelled in water at 50 °C. Differential scanning calorimetric (DSC) and compression tests at different temperatures revealed that poly(SA-*co*-MPGA) hydrogels with $x_{SA} > 0.5$ induce a crystalline-to-amorphous transition, which is a hard-to-soft transition at ~40 °C that is based on the formation/non-formation of a crystalline structure containing stearyl side chains. The hydrogels stored in water maintained an almost constant volume, independent of the temperature. The poly(SA-*co*-MPGA) hydrogel was soft, flexible, and deformed at 50 °C. However, the hydrogel stiffened when cooled to room temperature, and the deformation was reversible. The shape-memory function of poly(SA-*co*-MPGA) hydrogels is proposed for potential use in biomaterials; this is partially attributed to the use of MPGA, which consists of relatively biocompatible poly(ethylene glycol).

Keywords: thermosensitive hydrogel; shape-memory function; crystalline-to-amorphous transition; stearyl acrylate; biocompatible polymer

Citation: Tokuyama, H.; Iriki, R.; Kubota, M. Thermosensitive Shape-Memory Poly(stearyl acrylate-*co*-methoxy poly(ethylene glycol) acrylate) Hydrogels. *Gels* **2023**, *9*, 54. https://doi.org/10.3390/gels9010054

Academic Editors: Dong Zhang, Jintao Yang, Xiaoxia Le and Dianwen Song

Received: 20 December 2022
Revised: 4 January 2023
Accepted: 8 January 2023
Published: 10 January 2023

Copyright: © 2023 by the authors. Licensee MDPI, Basel, Switzerland. This article is an open access article distributed under the terms and conditions of the Creative Commons Attribution (CC BY) license (https://creativecommons.org/licenses/by/4.0/).

1. Introduction

Stimuli-sensitive hydrogels are promising candidates for use in soft actuators and intelligent biomaterials. Hydrogel actuators based on stimuli-sensitive polymers can alter their shape, size, or strength in response to external stimuli, such as heat, pH, light, and magnetic fields, resulting in flexible, complex mechanical motion and shape-memory function [1–3]. Robust mechanical and highly flexible properties are required for biomaterials such as artificial muscles, tendons, and ligaments. Hydrogels with excellent properties include slide-ring [4], double network [5], and tri-branched hydrogels [6].

Poly (*N*-isopropylacrylamide) (poly(NIPA)) is a polymer that is extensively used in hydrogel actuators. Poly(NIPA) is thermosensitive, with a lower critical solution temperature of ~33 °C in water [7,8]. Additionally, poly(NIPA) exhibits a hydrophilic/hydrophobic transition in response to temperature variation, and its hydrogel induces a volume phase transition. Thus, poly(NIPA)-based hydrogel actuators can stretch, shrink, bend, and twist [9–13]. However, its poor mechanical strength and thermosensitive volumetric changes may be disadvantageous for certain applications.

Poly(stearyl acrylate) (poly(SA)) is a thermosensitive polymer. Hydrophobic poly(SA) absorbs lipophilic solvents, but not water, and forms an organogel [14] instead of a hydrogel. Hydrogels consisting of poly(SA) were prepared by copolymerization of SA with a hydrophilic monomer. In the 1990s, Osada et al. [15–17] developed poly(SA-*co*-AA) (AA: acrylic acid) hydrogels that induced a crystalline-to-amorphous transition, which is an order-disorder transition associated with interactions between alkyl side chains at ~40 °C (depending on the monomer composition). This resulted in a significant change in the

Young's modulus and shape-memory function of the material. Furukawa et al. [18–21] developed poly(SA-*co*-DMAA) (DMAA: N,N-dimethylacrylamide) hydrogels for applications such as artificial lenses, bandages, and three-/four-dimensional (3D/4D) printing. Additionally, poly(SA-*co*-AM) (AM: acrylamide) was developed [22]. Thus, poly(SA)-based hydrogels induce a hard-to-soft transition in response to temperature while maintaining a constant volume.

In this study, the development and characterization of a novel poly(SA-*co*-MPGA) (MPGA: methoxy poly(ethylene glycol) acrylate) hydrogel as a potential thermosensitive biomaterial is reported. Notably, poly(ethylene glycol) (PEG) is a more biocompatible material than AA, AM, and DMAA; therefore, the poly(SA-*co*-MPGA) hydrogel is suitable for biomedical applications. Differential scanning calorimetric (DSC), compression, and shape-memory tests are conducted in relation to the crystalline-to-amorphous transition, as shown in Figure 1.

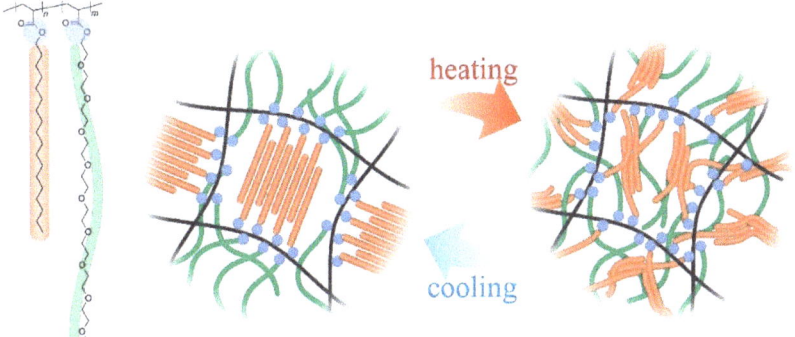

Figure 1. Chemical structure of poly(SA-*co*-MPGA) and an illustration of the crystalline-to-amorphous transition.

2. Results and Discussion

Figure 2 shows the swelling ratio of poly(SA-*co*-MPGA) hydrogels at 20–50 °C as a function of (a) temperature and (b) the mole fraction of SA, x_{SA}, in the pre-gel solution. Figure 2a shows the average values of the swelling ratio (that is, the size) of the hydrogels in the temperature range of 20–50 °C. The swelling ratio was almost constant and independent of the temperature. The swelling ratio increased with a decrease in x_{SA}, which corresponds to an increase in the mole fraction of hydrophilic MPGA. The hydrogels with x_{SA} <0.4 had a swelling ratio of >1, indicating a good water-swollen state. Hydrogels with x_{SA} >0.5 induced the crystalline-to-amorphous transition, as described later, and had a swelling ratio of <1.

The W/W_{dry} ratio was determined as a measure of the water absorption capacity of the gel, where W and W_{dry} were the masses of the hydrogel and dry gel, respectively. The W/W_{dry} for poly(SA-*co*-MPGA) with x_{SA} = 0.5 was 2.94 at 50 °C. For reference, the W/W_{dry} values reported in the literature were ~1.4 for poly(SA-*co*-DMAA) [21] and ~4.5 for poly(SA-*co*-AM) [22].

Figure 3 shows the DSC thermograms of the poly(SA-*co*-MPGA) hydrogels prepared with x_{SA} = 0.5, 0.7, and 0.8. These hydrogels had endothermic and exothermic peaks, whereas the hydrogel prepared with x_{SA} = 0.3 did not exhibit these peaks. A similar DSC thermogram for poly(SA) was reported in the literature [14,21]. Poly(SA) induces a crystalline-to-amorphous transition; the hydrophobic stearyl side chains form a crystalline structure at temperatures below the crystallization temperature T_c, and their packing becomes amorphous at temperatures above the melting temperature T_m. Previously, the T_m and T_c of the dry poly(SA) gel were reported to be 44.8 and 41.8 °C, respectively [14]. The DSC results demonstrated that the poly(SA-*co*-MPGA) hydrogel also induced a crystalline-to-amorphous transition (as shown in Figure 1). The T_c values were 43.3, 39.8, and 41.1 °C for poly(SA-co-MPGA) hydrogels prepared with x_{SA} = 0.5, 0.7, and 0.8, respectively. The T_m value was slightly unclear owing to the broad DSC peak, which was slightly higher than the T_c. The peak area for the poly(SA-co-MPGA) hydrogel prepared with x_{SA} = 0.5 was smaller than that of the hydrogels prepared with x_{SA} = 0.7 and 0.8, based on the amount of SA units per gram of hydrogel.

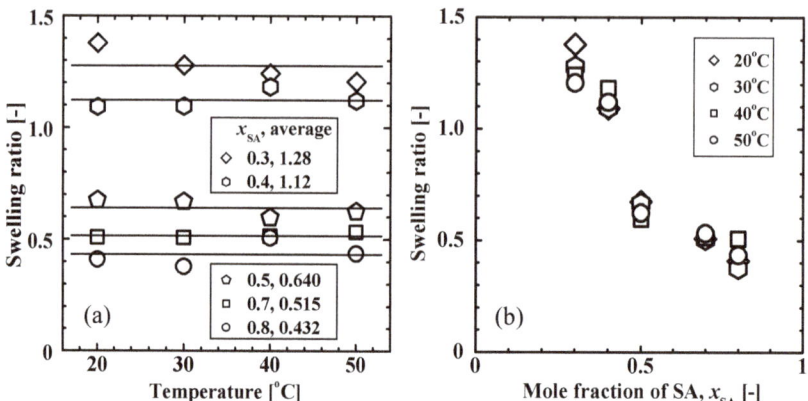

Figure 2. Swelling ratio of poly(SA-co-MPGA) hydrogels at 20–50 °C as a function of (**a**) temperature and (**b**) the mole fraction of SA, x_{SA}, in the pre-gel solution. The solid horizontal lines in (**a**) show the average swelling ratio at 20–50 °C.

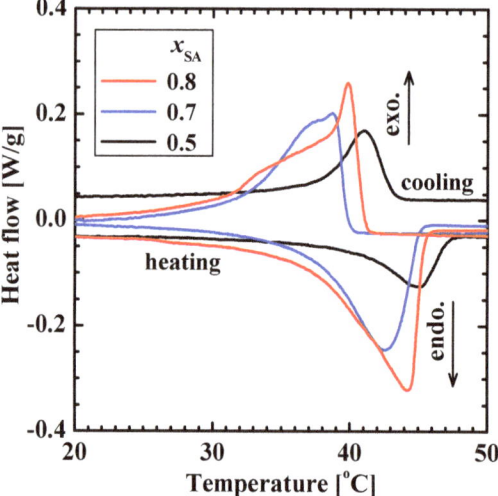

Figure 3. DSC thermograms of poly(SA-co-MPGA) hydrogels prepared with x_{SA} = 0.5, 0.7, and 0.8.

Figure 4 shows the compression test results of the cylindrical poly(SA-co-MPGA) hydrogels using a weight (36 g) as a load under various temperature conditions. The normalized length l/l_0 of the hydrogel prepared with x_{SA} = 0.7 was ~1 at 20–37.5 °C, and it decreased with an increase in temperature (>40 °C). This behavior indicates that the hydrogel stiffened <37.5 °C and softened >40 °C. The change in the hardness and softness, that is, the hard-to-soft transition, of the hydrogel was attributed to the crystalline-to-amorphous transition of the stearyl side chains of SA. The crystalline structure of the stearyl side chains function as pseudo-crosslinking points (as shown in Figure 1), enhancing the hydrogel strength. The hydrogel prepared with x_{SA} = 0.5 exhibited a similar thermosensitive behavior; however, the l/l_0 value was smaller at temperatures of >40 °C. The hydrogel prepared with x_{SA} = 0.3 had an l/l_0 value of ~0.8 at 20 °C, confirming its softness.

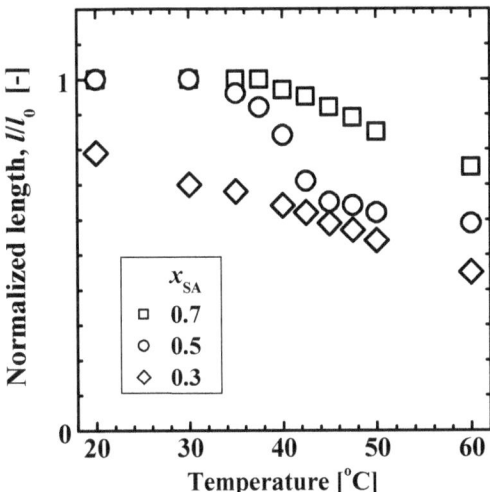

Figure 4. Normalized length, l/l_0, of cylinder-shaped poly(SA-co-MPGA) hydrogels with x_{SA} = 0.3, 0.5, and 0.7 as a function of temperature. l_0 is the initial length at 20 °C. l is the hydrogel length loaded with weight (36 g) at a given temperature.

The hydrogel strength at temperatures of >40 °C decreased with a decreasing x_{SA}. The hydrogel strength is primarily influenced by the swelling ratio shown in Figure 2. The rubber network theory, which was derived based on the statistical mechanics of crosslinked polymer networks, describes the relationship $\tau \propto (\nu_e \, \phi_p^{-2/3})$ in the stress-strain curves of the tensile or compressive strengths of gels, where τ is the stress required for a given deformation, ν_e is the effective crosslinking density, and ϕ_p is the volume fraction of the polymer in the hydrogel [23–25]. As shown in Figure 2, a decrease in x_{SA} causes an increase in the swelling ratio, resulting in reductions in ν_e and ϕ_p and consequently a decrease in τ.

Figure 5 shows the shape memory function of the poly(SA-co-MPGA) hydrogel prepared with x_{SA} = 0.5. Initially, the hydrogel was rod-shaped. The hydrogel was soft and flexible at 50 °C and deformed into an S shape. When the hydrogel was cooled to room temperature, it stiffened, and the S shape was fixed. When the hydrogel was heated to 50 °C, it reverted to its original shape. The deformation based on the hard-to-soft transition induced by heating or cooling occurs within a few minutes and repeatedly. The S-shaped hydrogel structure can be retained indefinitely in water at room temperature, which was confirmed for several months.

The poly(SA-co-MPGA) hydrogel deforms at a temperature slightly higher than body temperature and stiffens at body temperature. Thus, the poly(SA-co-MPGA) hydrogel is proposed for use as a cast-like, anti-adhesive, or stent material with a well-fitted shape to reinforce and protect injured or post-surgery organs and tissues in the body.

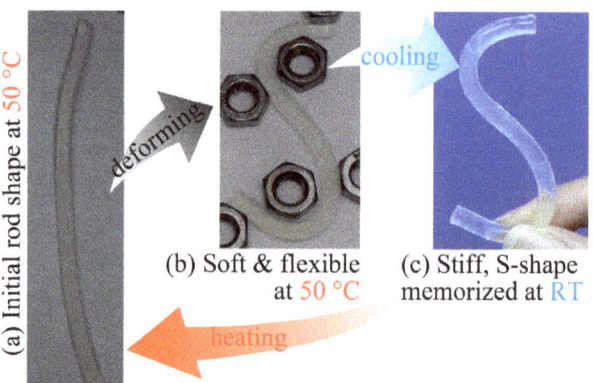

Figure 5. Photographs of the shape-memory function of poly(SA-*co*-MPGA) hydrogel with x_{SA} = 0.5: (**a**,**b**) in water at 50 °C and (**c**) at room temperature (~20 °C).

3. Conclusions

Poly(SA-*co*-MPGA) hydrogels were prepared by free-radical copolymerization of SA, MPGA, and EGDM in ethanol, followed by washing, drying, and re-swelling in water at 50 °C. The DSC and compression tests performed at different temperature conditions revealed that poly(SA-*co*-MPGA) hydrogels with x_{SA} >0.5 induce a crystalline-to-amorphous transition, which is a hard-to-soft transition that occurs at ~40 °C. The hydrogels had an almost constant volume, independently of the temperature. The shape-memory function of poly(SA-*co*-MPGA) hydrogel is that it is soft, flexible, and deformed at temperatures of >40 °C and that it stiffens when cooled to <37.5 °C. Additionally, the deformation of hydrogel is reversible.

4. Materials and Methods

4.1. Preparation of Poly(SA-co-MPGA) Gels

Copolymer gels with various concentrations of SA and MPGA (average molecular weight: 483) were synthesized by free-radical polymerization. Ethanol was used as a solvent to dissolve hydrophobic SA and hydrophilic MPGA. The monomer solution contained SA, MPGA, ethylene glycol dimethacrylate (EGDM; crosslinking monomer), and N,N,N′,N′-tetramethylethylenediamine (TEMED; polymerization accelerator). The initiator solution contained 2,2′-azobis(2,4-dimetylvaleronitrile) (ADVN; polymerization initiator). Nitrogen gas was bubbled through each solution for 1 h to remove dissolved oxygen. Subsequently, the initiator solution was added to the monomer solution in a polytetrafluoroethylene (PTFE) tube (inner diameter: 6 mm). Polymerization was performed at 60 °C for 1 d in a nitrogen atmosphere. The overall concentration of the primary monomers in the pre-gel solution was 1000 mol/m^3; for example, 700 mol/m^3 of SA and 300 mol/m^3 of MPGA, corresponding to an SA mole fraction, x_{SA}, of 0.7. The concentrations of EGDM, TEMED, and ADVN were 100, 30, and 20 mol/m^3, respectively, for all the gels. The resulting gels were cut into cylinders with a length of 6 mm. Subsequently, the gels were washed with ethanol at 50 °C to remove non-crosslinked chemicals and then dried in an oven at 50 °C.

4.2. Swelling Properties in Water

The dry, cylindrical gel was immersed in water at 50 °C for several days, and water was absorbed to obtain the poly(SA-*co*-MPGA) hydrogel. The hydrogel diameter, d [mm], at swelling equilibrium was measured using a photograph taken with a digital camera. Subsequently, the hydrogel diameter was measured at 40, 30, and 20 °C. The swelling ratio was defined as the hydrogel volume divided by the volume of the as-synthesized gel, and was calculated as follows: $(d/6)^3$.

4.3. Compression Test

The temperature dependence of the softness and hardness of the poly(SA-co-MPGA) hydrogel was evaluated. The cylindrical hydrogel was vertically placed in a glass test tube, and water was added to half the height of the hydrogel. Subsequently, the test tube was placed in a constant-temperature water bath at 20 °C. The initial length l0 of the hydrogel at 20 °C was measured using a digital camera. A total weight (36 g) was placed on the hydrogel. Subsequently, the length l of the hydrogel was measured after several minutes. Under a continuous load, the temperature was increased stepwise, and the hydrogel length was measured at each temperature.

4.4. DSC Analysis

A differential scanning calorimeter (DSC-60, Shimadzu Co., Kyoto, Japan) was used to perform DSC analysis. The poly (SA-co-MPGA) hydrogel was ground, and the ground sample (3.2 mg) was enclosed in an aluminum cell. α-Alumina was used as a reference material and enclosed in another cell. The cells were placed in a sample chamber under nitrogen gas flow. For DSC measurements, the cells were heated and subsequently cooled between 0 and 60 °C at a rate of 2 °C/min.

Author Contributions: Conceptualization, H.T.; methodology, H.T., R.I. and M.K.; software, H.T., R.I. and M.K.; validation, H.T., R.I. and M.K.; formal analysis, H.T., R.I. and M.K.; investigation, H.T., R.I. and M.K.; resources, H.T., R.I. and M.K.; data curation, H.T., R.I. and M.K.; writing—original draft preparation, H.T., R.I. and M.K.; writing—review and editing, H.T.; visualization, H.T.; supervision, H.T.; project administration, H.T.; funding acquisition, H.T. All authors have read and agreed to the published version of the manuscript.

Funding: This research received no external funding.

Data Availability Statement: Not applicable.

Acknowledgments: MPGA was generously supplied by Kyoeisha Chemical Co., Ltd.

Conflicts of Interest: The authors declare no conflict of interest.

References

1. Apsite, I.; Salehi, S.; Ionov, L. Materials for smart soft actuator systems. *Chem. Rev.* **2022**, *122*, 1349–1415. [CrossRef] [PubMed]
2. Kim, J.; Kim, J.W.; Kim, H.C.; Zhai, L.; Ko, H.U.; Muthoka, R.M. Review of soft actuator materials. *Int. J. Precis. Eng. Manuf.* **2019**, *20*, 2221–2241. [CrossRef]
3. Han, I.K.; Chung, T.; Han, J.; Kim, Y.S. Nanocomposite hydrogel actuators hybridized with various dimensional nanomaterials for stimuli responsiveness enhancement. *Nano Converg.* **2019**, *6*, 18. [CrossRef] [PubMed]
4. Liu, C.; Morimoto, N.; Jiang, L.; Kawahara, S.; Noritomi, T.; Yokoyama, H.; Mayumi, K.; Ito, K. Tough hydrogels with rapid self-reinforcement. *Science* **2021**, *372*, 1078–1081. [CrossRef]
5. Nonoyama, T.; Gong, J.P. Tough double network hydrogel and its biomedical applications. *Annu. Rev. Chem. Biomol. Eng.* **2021**, *12*, 393–410. [CrossRef] [PubMed]
6. Fujiyabu, T.; Sakumichi, N.; Katashima, T.; Liu, C.; Mayumi, K.; Chung, U.I.; Sakai, T. Tri-branched gels: Rubbery materials with the lowest branching factor approach the ideal elastic limit. *Sci. Adv.* **2022**, *8*, eabk0010. [CrossRef]
7. Hirokawa, Y.; Tanaka, T. Volume phase transition in a nonionic gel. *J. Chem. Phys.* **1984**, *81*, 6379–6380. [CrossRef]
8. Tokuyama, H.; Mori, H.; Hamaguchi, R.; Kato, G. Prediction of the lower critical solution temperature of poly(N-isopropylacrylamide-co-methoxy triethyleneglycol acrylate) in aqueous salt solutions using support vector regression. *Chem. Eng. Sci.* **2021**, *231*, 116325. [CrossRef]
9. Deng, K.; Rohn, M.; Gerlach, G. Design, simulation and characterization of hydrogel-based thermal actuators. *Sens. Actuators B* **2016**, *236*, 900–908. [CrossRef]
10. Warren, H.; Shepherd, D.J.; in het Panhuis, M.; Officer, D.L.; Spinks, G.M. Porous PNIPAm hydrogels: Overcoming diffusion-governed hydrogel actuation. *Sens. Actuators A* **2020**, *301*, 111784. [CrossRef]
11. Choi, J.G.; Spinks, G.M.; Kim, S.J. Mode shifting shape memory polymer and hydrogel composite fiber actuators for soft robots. *Sens. Actuators A* **2022**, *342*, 113619. [CrossRef]
12. Liu, J.; Jiang, L.; Liu, A.; He, S.; Shao, W. Ultrafast thermo-responsive bilayer hydrogel actuator assisted by hydrogel microspheres. *Sens. Actuators B* **2022**, *357*, 131434. [CrossRef]
13. Tokuyama, H.; Sasaki, M.; Sakohara, S. Preparation of a novel composition-gradient thermosensitive gel. *Colloids Surf. A* **2006**, *273*, 70–74. [CrossRef]

14. Tokuyama, H.; Kato, Y. Preparation of thermosensitive polymeric organogels and their drug release behaviors. *Eur. Polym. J.* **2010**, *46*, 277–282. [CrossRef]
15. Matsuda, A.; Sato, J.; Yasunaga, H.; Osada, Y. Order-disorder transition of a hydrogel containing an N-alkyl acrylate. *Macromolecules* **1994**, *27*, 7695–7698. [CrossRef]
16. Osada, Y.; Matsuda, A. Shape memory in hydrogels. *Nature* **1995**, *376*, 219. [CrossRef]
17. Kagami, Y.; Gong, J.P.; Osada, Y. Shape memory behaviors of crosslinked copolymers containing stearyl acrylate. *Macromol. Rapid Commun.* **1996**, *17*, 539–543. [CrossRef]
18. Hasnat Kabir, M.; Gong, J.; Watanabe, Y.; Makino, M.; Furukawa, H. Hard-to-soft transition of transparent shape memory gels and the first observation of their critical temperature studied with scanning microscopic light scattering. *Mater. Lett.* **2013**, *108*, 239–242. [CrossRef]
19. Hasnat Kabir, M.; Hazama, T.; Watanabe, Y.; Gong, J.; Murase, K.; Sunada, T.; Furukawa, H. Smart hydrogel with shape memory for biomedical applications. *J. Taiwan Inst. Chem. Eng.* **2014**, *45*, 3134–3138. [CrossRef]
20. Shiblee, M.D.N.I.; Ahmed, K.; Yamazaki, Y.; Kawakami, M.; Furukawa, H. Light scattering and rheological studies of 3D/4D printable shape memory gels based on poly (N,N-dimethylacrylamide-*co*-stearyl acrylate and/or lauryl acrylates). *Polymers* **2021**, *13*, 128. [CrossRef]
21. Kabir, M.H.; Ahmed, K.; Furukawa, H. The effect of cross-linker concentration on the physical properties of poly(dimethyl acrylamide-*co*-stearyl acrylate)-based shape memory hydrogels. *Microelectron. Eng.* **2016**, *150*, 43–46. [CrossRef]
22. Lin, X.K.; Chen, L.; Zhao, Y.P.; Dong, Z.Z. Synthesis and characterization of thermoresponsive shape-memory poly(stearyl acrylate-*co*-acrylamide) hydrogels. *J. Mater. Sci.* **2010**, *45*, 2703–2707. [CrossRef]
23. Flory, P.J.; Rehner, J., Jr. Statistical mechanics of cross–linked polymer networks II. Swelling. *J. Chem. Phys.* **1943**, *11*, 521–526. [CrossRef]
24. James, H.M.; Guth, E. Theory of the elastic properties of rubber. *J. Chem. Phys.* **1943**, *11*, 455–481. [CrossRef]
25. Tokuyama, H.; Nakahata, Y.; Ban, T. Diffusion coefficient of solute in heterogeneous and macroporous hydrogels and its correlation with the effective crosslinking density. *J. Membr. Sci.* **2020**, *595*, 117533. [CrossRef]

Disclaimer/Publisher's Note: The statements, opinions and data contained in all publications are solely those of the individual author(s) and contributor(s) and not of MDPI and/or the editor(s). MDPI and/or the editor(s) disclaim responsibility for any injury to people or property resulting from any ideas, methods, instructions or products referred to in the content.

Hydrolytic Stability of Crosslinked, Highly Alkaline Diallyldimethylammonium Hydroxide Hydrogels

Tim B. Mrohs and Oliver Weichold *

Institute of Building Materials Research, RWTH Aachen University, Schinkelstraße 3, 52062 Aachen, Germany
* Correspondence: weichold@ibac.rwth-aachen.de

Abstract: The aim of this study was to evaluate the persistence of alkaline hydrogels based on a common (N,N'-methylenebisacrylamide, BIS) and three recently published tetraallyl crosslinkers. Such hydrogels have been shown to be suitable materials for the rehabilitation of cementitious materials. Of the four crosslinkers under investigation, N,N,N',N'-tetraallylpiperazinium dibromide decomposed quickly in 1 M KOH solution and was not considered further. BIS showed the first signs of a decomposition after several days, while tetraallylammonium bromide and N,N,N',N'-tetraallyltrimethylene dipiperidine dibromide remained unaffected. In contrast to BIS, which suffers from low solubility in water, the two tetraallyl crosslinkers show unlimited miscibility with diallyldimethylammonium hydroxide solutions. For the study, gels with up to 50 wt % crosslinker were prepared. Of these, gels containing tetraallylammonium bromide always show the highest degrees of swelling, with a peak value of 397 g/g at a content of 2 wt %. Under accelerated ageing at 60 °C for 28 d, gels crosslinked with BIS ultimately turned liquid, while the storage modulus and the degree of swelling of the two tetraallyl-crosslinked gels remained unchanged. This indicates that alkaline gels can be suitable for long application periods, which are common for rehabilitation measures in the construction industry.

Keywords: hydrogel; copolymer; durability; hydrolysis; swelling; rheology; crosslinker

1. Introduction

Hydrogels have become increasingly important in research and industry in recent years. Gels in general are crosslinked polymer networks, which can absorb and release various liquids without losing their discrete three-dimensional structure. If the absorbed medium is water, the polymer network is referred to as hydrogel [1]. The major advantage of hydrogels is the potential to specifically tailor the chemical structure, which allows for a wide range of applications. Non-ionic hydrogels, for example, are often used in protein analysis [2] or biomedical applications [3] due to their pH-independent swelling properties and their insensitivity to salt concentrations [4]. However, the majority of hydrogels are ionic, such as the well-known poly(sodium acrylate), which is used as a superabsorber for diapers [5] or as a shrinkage-reducing agent in concrete [6]. Ionic hydrogels usually exhibit significantly higher degrees of swelling and can respond to changes in pH value and/or salt concentration in the surrounding medium [7].

Hydrogels that are particularly suitable for the construction industry can be obtained by using cationic networks with hydroxide as a counterion; these have recently been realized based on diallyldimethylammonium hydroxide (DADMAOH) as a monomer and N,N'-methylenebisacrylamide (BIS) as a crosslinker. Such highly alkaline polymer networks were not only shown to be valuable materials in the rehabilitation of steel-reinforced concrete by exchanging carbonate ions in aged concrete with hydroxide ions, thereby restoring the high pH value necessary for preventing steel corrosion [8], but also as a coupling material for electrochemical chloride extraction [9]. A similar hydrogel was recently used to seal water-bearing cracks, while at the same time restoring the protective passive-layer on the exposed parts of steel rebars [10]. Although the system has already been tested successfully in field trials [9], the crosslinker BIS appears to be a weak point for a number of reasons: BIS suffers from a rather low solubility in water (approx. 20 g/L at 20 °C [11]), which limits the possibility to prepare firm gels. Moreover, as an acrylate derivative, BIS polymerizes significantly faster than the diallyl (DADMA$^+$) unit, which was shown to lead to inhomogeneous networks [12]. Gels crosslinked with BIS also made the qualitative impression of softening over the course of several months. The hydrolysis of the bisamide liberates formaldehyde, which is undesirable in large-scale applications [13].

We have recently reported on the synthesis of three new tetraallylammonium-based crosslinkers, namely tetraallylammonium bromide, N,N,N',N'-tetraallylpiperazinium dibromide, and N,N,N',N'-tetraallyltrimethylene dipiperidine dibromide [12], and we used these to successfully crosslink diallyldimethylammonium chloride, a pH-neutral derivative of DADMAOH. Due to their better solubility in water, a wider range of crosslinking densities can be obtained, and due to their structural similarity, the copolymerisation leads to homogeneous networks [12]. The question now arises as to how these crosslinkers perform in the alkaline diallyldimethylammonium hydroxide (DADMAOH) system designed for application in, e.g., cementitious materials. This question is addressed by first evaluating the persistence of the pure compounds in alkaline media and then validating these findings by monitoring the rheological and swelling properties of gels under accelerated ageing. The results are compared to gels crosslinked with BIS.

2. Results and Discussion

The copolymerization of DADMAOH with the crosslinkers **1a–c** follows the published procedure for the polymerization of DADMAOH with BIS using a redox initiation system consisting of potassium peroxodisulfate and sodium disulfite [8]. As shown in Figure 1, the diallyldimethylammonium unit polymerizes under ring closure, triggered by the attack of a radical. The same mechanism also operates in the crosslinkers **1a–c**. BIS, on the other hand, polymerises by linear radical addition to each of the double bonds—i.e., it does not form a ring—and, therefore, exhibits clearly different copolymerization characteristics than **1a–c** with DADMA$^+$ monomers [12].

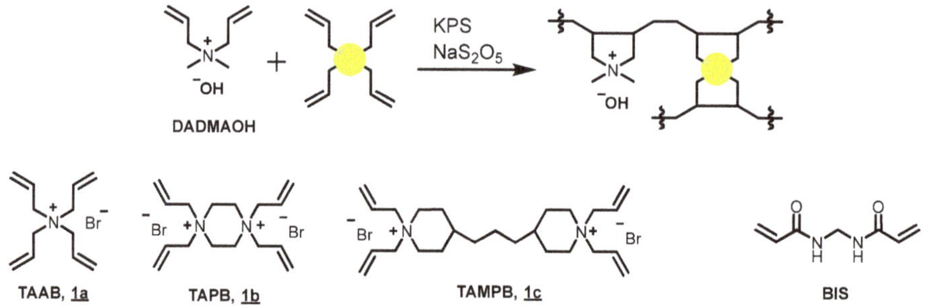

Figure 1. Top: general scheme for the crosslinking polymerization of DADMAOH; bottom: structures of the cross-linkers **1a–c** and BIS used in this study.

Initial tests for the preparation of such highly alkaline DADMAOH gels using the tetraallyl crosslinkers **1a–c** appeared only successful with TAAB (**1a**) and TAMPB (**1c**). The mixture with TAPB (**1b**) turned from colourless to yellow and dark orange in a short period of time without forming a gel and simultaneously developed a strong fish-like odour, indicating the release of amines. This is even more surprising, since poly(acrylate)s crosslinked with TAPB (**1b**) were found to be largely unaffected by boiling in NaOH solutions [14]. Therefore, control experiments regarding the hydrolytic stability of the crosslinkers **1a–c** and the previously used BIS were run by monitoring the ^{1}H-NMR spectra in D$_2$O containing 1 mol/L KOH over the course of seven days (Figure 2). The spectra in pure D$_2$O were used as a reference.

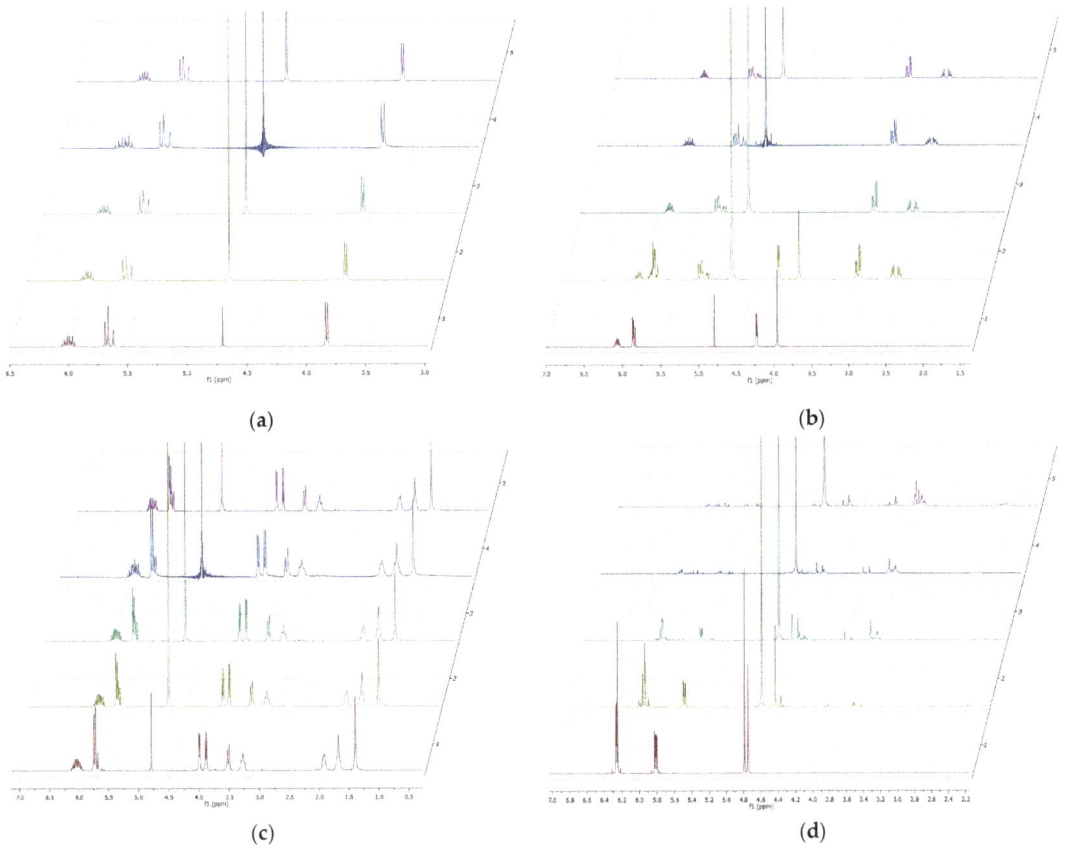

Figure 2. ^{1}H-NMR spectra of the pure crosslinkers in 1 M KOH solution in D$_2$O over the course of 7 days: (**a**) tetraallyl ammonium bromide, (**b**) N,N,N',N'-tetraallyl piperazinium dibromide, (**c**) N,N,N',N'-tetraallyl trimethylene dipiperidine dibromide, (**d**) BIS. The numbers on the right axis indicate the time code: (1) reference spectrum without KOH, (2) 1 M KOH in D$_2$O on the same day, (3) after 24 h, (4) after 48 h, (5) after 7 days.

The control experiments showed no change in the appearance of the ^{1}H-NMR spectra of TAAB (Figure 2a) and TAMPB (Figure 2c) over the period of 7 days. It can, thus, be concluded that these two are resistant to alkaline hydrolysis under these conditions. BIS, on the other hand, already shows (Figure 2d) the first signals of decomposition on the same day (Figure 2d(2)). Their intensity increases over time, while simultaneously the characteristic signals of BIS at 6.25 ppm, 5.83 ppm, and 4.75 ppm decrease and are completely lost after 48 h. In the case of TAPB (Figure 2b), signals of the crosslinker at 6.06 ppm, 5.86 ppm, 4.23 ppm, and 3.96 ppm could be identified on the same day in KOH/D$_2$O, but these disappeared completely after 24 h (Figure 2b(3)). The susceptibility of TAPB to decomposition in alkaline media could be explained by a Hofmann-type elimination (Figure 3). The mechanism is particularly favoured by the proximity of the two positive charges on the piperazine ring, which creates tension within the ring due to the electrostatic repulsion and renders the hydrogen atoms α to the positive charges in **1b** more acidic. The latter can be indirectly observed by comparing the position of these H atoms (4.7 ppm, Figure 2b) to those in compound **1c** (3.7 ppm, Figure 2c), which are shifted to a higher field. The decrease of the signal intensity in the NMR is accompanied by the formation of a water-insoluble phase, which deposits a supernatant layer and appears to contain various decomposition products such as allylpiperazine, diallylvinylamine, and diallylamine, amongst others. An ^{1}H-NMR spectrum of this in CDCl$_3$ is given in Figure S1 (Supplementary Material).

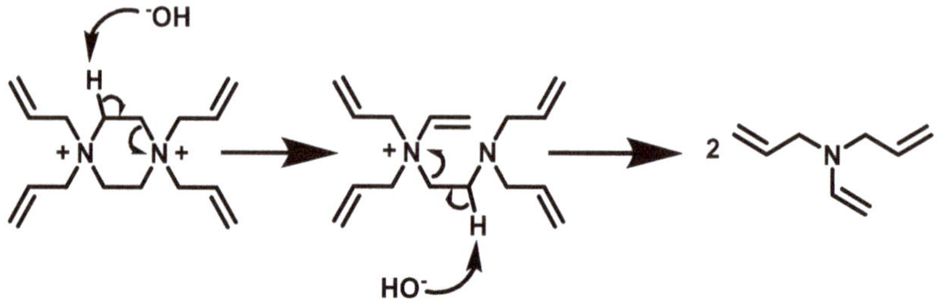

Figure 3. Potential beginning of the alkaline degradation of TAPB (**1b**) by Hofman-type elimination.

As a result, TAPB (**1b**) is considered unsuitable as a crosslinker in the highly alkaline media and will therefore not be considered further. BIS, on the other hand, decomposes much more slowly and forms at least stable gels. BIS has previously been used to crosslink highly alkaline gels [8–10] and will therefore be used as reference for the allyl crosslinkers **1a** and **1c**.

In order to find a suitable degree of crosslinking for the durability tests, the swelling properties of the gels were first determined as a function of the crosslinker content. For this purpose, DADMAOH hydrogels containing 2–50 mol% TAAB (**1a**) or TAMPB (**1c**) were prepared (Figure 4). In the case of BIS, the gels could only be prepared with 3 and 4 mol%, as amounts of less than 3 mol% did not result in stable gels, and 4 mol% is the solubility limit of BIS in this mixture. Technically, higher BIS/monomer ratios are possible at lower monomer concentrations, but such gels again exhibit poor mechanical stability. All gels were polymerized over a period of 3 weeks to ensure complete conversion. The comparatively long polymerization times are based on those previously observed for diallyldimethylammonium chloride gels [12].

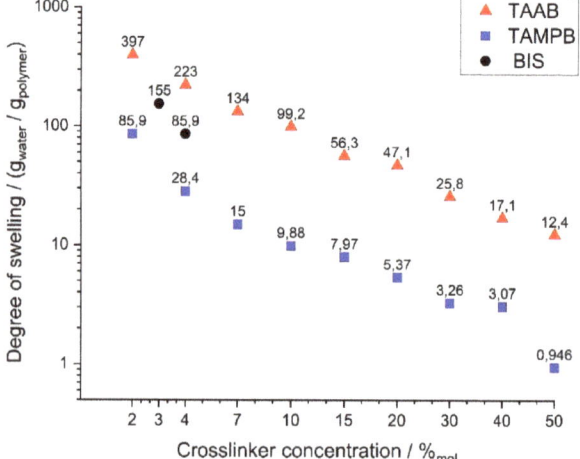

Figure 4. Swelling behaviour of poly(DADMAOH) gels with different crosslinkers as a function of the crosslinker concentration in bidistilled water. Reproducibility is approx. ±10–14%. Due to the logarithmic scale, the error bars are smaller than the symbols and are therefore omitted.

Three trends can be observed from Figure 4: (i) as expected, the degree of swelling decreases for all gels with increasing crosslinker content, (ii) this seems to be less pronounced for TAAB (**1a**) than for TAMPB (**1c**), as the ratio of the degrees of swelling increases from 4.6 at 2% to 10 at 10% and to 13.1 at 50%, and (iii) TAAB (**1a**) consistently results in by far the highest degrees of swelling, with a measurable value of up to 397 g/g at a content of 2%. These values are comparable to common acrylate superabsorbent polymers [15,16]. BIS, on the other hand, is not only limited by the solubility but also by a seemingly much stronger decrease in the degree of swelling. Thus, the application range of gels with crosslinkers **1a,c** is much broader than that of the original BIS-DADMAOH system.

As stated above, DADMAOH crosslinked with 2% BIS does not form stable gels. To compensate this, methacrylamide (MAA) has previously been added as comonomer, which stabilizes the resulting gels due to the formation of hydrogen bonds and dipole-charge interactions [8]. As a side effect, the gels also become more pliable, which has a favourable effect on the processability. Although the addition of a comonomer to obtain coherent gels at low crosslinker contents is not necessary when using the tetraallyl compounds **1a,c**, it was interesting to test the effect of MAA on the present system in view of potential later applications. For this, the (chemical) crosslinker content was fixed at 10 mol% **1a,c**, and the MAA content was increased from 0 to 8 and then to 20% molar fraction; i.e., the ratio of crosslinker to the total amount of monomers was equal in all mixtures. After complete polymerization, the storage modulus and swelling properties of the resulting gels were analysed (Figure 5).

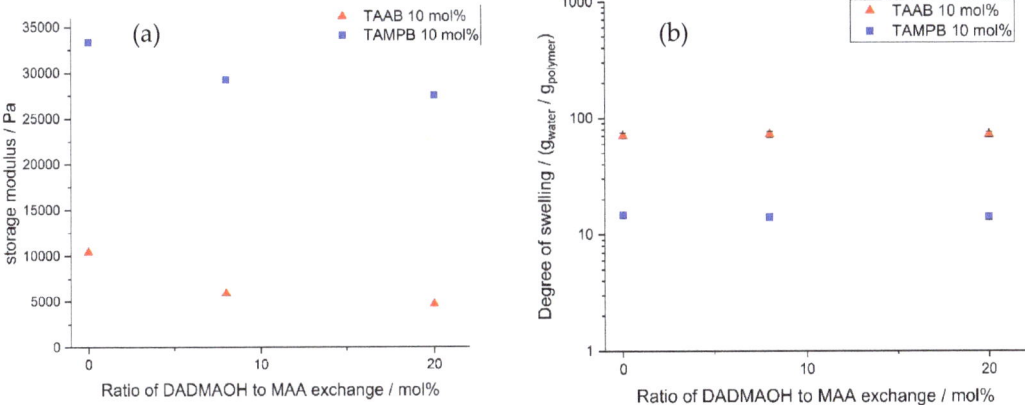

Figure 5. Storage modulus (**a**) and equilibrium degree of swelling (**b**) of gels crosslinked with 10 mol% TAAB (triangles) and TAMPB (squares), as a function of the molar fraction of methacrylamide. The storage modulus was determined at an amplitude of 1%, a frequency of 1 Hz, and a contact pressure of 5 N.

Figure 5a shows that for both TAAB (**1a**) and TAMPB (**1c**), the storage modulus decreases with increasing MAA content. This is in contrast to previous studies using 2 mol% BIS as a crosslinker, which noted a stiffening of the gels upon increasing MAA content [8]. The rate of decrease appears similar for both crosslinkers, despite the initial large difference in storage modulus (33.3 kPa for TAMPB, 10.4 kPa for TAAB). This leads to the assumption that the decrease is independent of the molecular structure. The decrease in storage moduli could, therefore, originate in a combination of two effects: gels crosslinked with tetraallyl compounds exhibit a homogeneous distribution of nodes, and at 10 mol% crosslinker, the gel appears to be too rigid for the weaker hydrogen bonds and dipole-charge interactions to be noticeable. On the other hand, the uncharged monomer reduces the charge density in the chains and reduces the electrostatic repulsion. This renders the chains more flexible. However, the reduction of the charge density does not seem to affect the swelling properties (Figure 5b). MAA was, therefore, used in the following experiments.

For the rheological investigations, a crosslinker content of 2 mol% was selected for the TAMPB and TAAB gels and 4 mol% for the BIS gels due to the otherwise insufficient gel stability. In addition, 8 mol% methacrylamide based on DADMAOH was added to the polymerization solutions. Initially, all gels were cured in individual vials for 3 weeks and then 3–5 mm thick slices were cut from the centre of each gel block. The measurement errors due to the thickness variation of the gel slices are not significant here [17]. In order to be able to carry out the measurements reliably, the upper plate needs to contact the gel completely. This is not an issue in the case of very soft gels such as the ones crosslinked with BIS. For these heterogeneous gels, the dependence of storage modulus and normal force is not linear due to the macroporous structure, so measurements should be taken at low normal force [18]. Gels based on the tetraallyl crosslinkers **1a,c** appear firmer, despite the lower crosslinker content. Here, an additional pressure in the form of a normal force was needed to achieve full contact. Since it is known that this has a considerable influence on the determined moduli [17], the effect of the normal force on the present system was tested using a gel crosslinked with 2 mol% TAMPB (**1c**) at an amplitude of 1% (Figure 6).

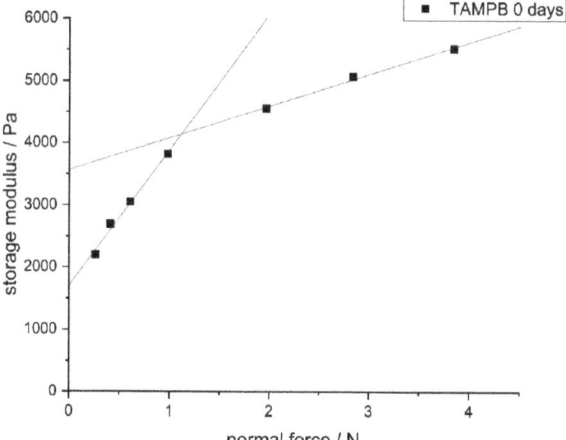

Figure 6. Storage modulus of a poly(DADMAOH) hydrogel with 2 mol% TAMPB and 8 mol% methacrylamid as a function of the applied normal force. The values were recorded at an amplitude of 1% and a frequency of 1 Hz 4 weeks after starting the polymerization.

Figure 6 shows that the observed storage modulus increases with increasing normal force, but with two different dependencies. At forces < 1 N, a very sharp increase can be seen, which changes to a significantly lower slope at forces > 1 N. From visual observation, the initial sharp increase can be addressed to the increasing contact area of the gels with the plates of the rheometer. In accordance with the literature, the subsequent region with lower slope is the result of the polymer chains in rigid gels being compressed, which results in macroscopic stiffening [17]. In order to remove the first effect, all samples crosslinked with TAAB (**1a**) and TAMPB (**1c**) were analysed using a normal force of 1 N. This was in agreement with the observations in the work of Karpushkin [18], where the values of approx. 0.1 to 1 N were reported. For significantly harder gels, it is necessary to use higher contact pressures, but these are determined by the same procedure in the later progress of the work.

To assess the susceptibility of the gels to alkaline hydrolysis, the polymerisation was allowed to continue for 3 weeks. The reference values ($t = 0$ in Figure 7) were determined at this point. The gels were then stored at room temperature and also at 60 °C and continuously monitored by analysing their rheological and swelling properties. At room temperature, a very small decrease in the storage modulus of BIS-crosslinked DADMAOH gels was observed, but not in those gels containing the tetraallyl-crosslinkers **1a,c**. The same applies to the degree of swelling at room temperature (Figures S2 and S3 in the Supplementary Materials). Therefore, the experiments were repeated at 60 °C to accelerate potential decomposition reactions (Figure 7).

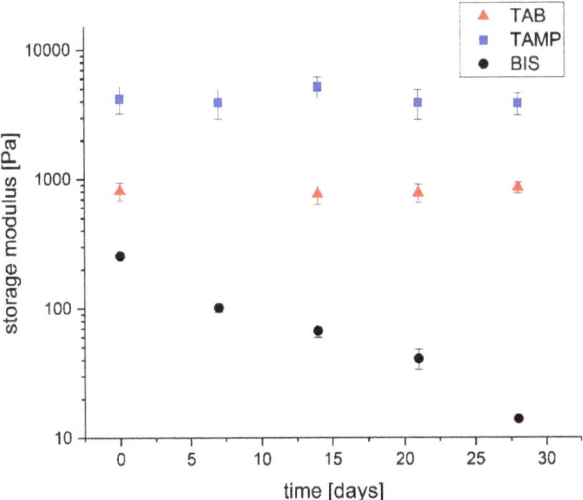

Figure 7. Change of the storage moduli of poly(DADMAOH-*co*-MAA)-hydrogels crosslinked with 2 mol% TAAB (triangles) or TAMPB (squares) and 4 mol% BIS (circles) at 60 °C.

Over the course of 28 days at 60 °C, the storage modulus of the BIS-crosslinked samples continuously decreased from 256 Pa to approx. 13 Pa. This enormous loss can also be observed haptically and visually, since after 28 days the gels were fluid. In contrast, the storage modulus of gels crosslinked with TAAB (**1a**) or TAMPB (**1c**) appeared to be constant. To corroborate this, a linear regression of the values in Figure 7 afforded a slope of -0.04 ± 0.0009 for gels crosslinked with BIS, while for the other two, the slope was 0 within the scatter of the measured values. On the molecular scale, the liquefaction can be explained by a degradation of the crosslinking points. This is a strong indication that gels crosslinked with **1a,c** are resistant to alkaline hydrolysis over the period of observation (28 d, 60 °C).

In order to verify the above results, the swelling properties of the samples in Figure 7 were also determined (Figure 8).

Figure 8 shows that for TAAB-crosslinked gels, the degree of swelling is very much constant at approx. 250 g/g over the course of 28 days at 60 °C. The same applies to TAMPB-crosslinked gels, albeit with a lower degree of swelling of approx. 70 g/g. Since swelling mainly depends on the crosslinking density within the gels, it can be concluded that the crosslinking points persist in the highly alkaline environment. This confirms the findings in Figure 7. In contrast, gels crosslinked with BIS are again strongly affected, but despite the continuous decrease of the storage modulus shown in Figure 7, the degree of swelling increases at first from approx. 120 g/g to approx. 180 g/g. After 14 days, it decreases at an increasing rate and cannot be determined after 28 d since the sample is liquid. This also supports the decomposition of BIS already suspected in Figure 7. The initial increase is also, in our opinion, evidence for the beginning decomposition of the crosslinking points, since a lower crosslinking density gives—within certain limits—rise to higher degrees of swelling. The turning point over the course of the swelling curve is also initiated by parts of the gel flowing through the 90 μm wide meshes of the polyester teabag. As gels crosslinked with BIS have previously been shown to possess a non-uniform network structure [12], small amounts of hydrolysis might sever larger portions of polymer from the gel, which are then lost in the experiment.

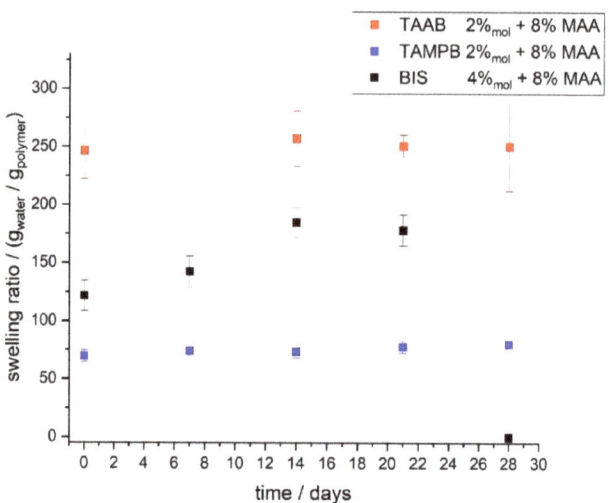

Figure 8. Equilibrium degree of swelling of poly(DADMAOH-co-MAA) hydrogels crosslinked with 2 mol% TAAB/2 mol% TAMPB/4 mol% BIS at different times after storage of the polymerized gels at 60 °C. All gels were lyophilized before the swelling tests.

3. Conclusions

The persistence of the present highly alkaline hydrogels depends on the nature of the crosslinker. The amide bonds in the commonly used N,N'-methylenebisacrylamide are subject to slow hydrolysis, which ultimately causes liquefication of the gels. Rather unexpectedly, N,N,N',N'-tetraallylpiperazinium dibromide decomposes so quickly that alkaline gels do not form. The reason could be the proximity of the two positive charges in the six-membered ring, which renders the α-H atoms strongly acidic. This favours a Hofmann-type elimination. Gels crosslinked with tetraallylammonium bromide and N,N,N',N'-tetraallyltrimethylene dipiperidine dibromide did not show any signs of decomposition after 28 d at 60 °C, which translates to at least 15 months at ambient temperature. This is sufficient for basically all relevant rehabilitation measures, such as realkalisation or chloride extraction. Both suitable tetraallyl crosslinkers allow the preparation of gels with a wide range of mechanical and swelling properties and are compatible with methacrylamide as comonomer, which renders the gels more pliable.

4. Materials and Methods

4.1. Materials

Diallyldimethylammonium chloride (65 wt % in H_2O) and triallylamine (99%) were purchased from Sigma Aldrich, and piperazine and D_2O (99.9%) were obtained from Merck KGaA (Darmstadt, Germany). Allyl bromide (99%), potassium carbonate (99%), methanol (99%), and 1,3-bis(4-piperidyl)propane (97+%) were purchased from Alfa Aesar (Kandel, Germany). The anion exchange-resin Lewatit Monoplus MP 800 was provided by Lanxess (Leverkusen, Germany). Chloroform (≥99%), acetone (≥99%), bidistilled water, potassium hydroxide, dichloromethane (≥99%), potassium persulfate, sodium metabisulphite, N,N'-methylenebisacrylamide and sodium hydroxide (97%) were obtained from VWR International GmbH (Darmstadt, Germany). All chemicals were used as received. Polyester filter bags with a maximum mesh size of 90 μm were purchased from Rosin Tech Products (Bethpage, NY, USA).

4.2. Synthesis of the Crosslinkers

Tetraallylammonium bromide **1a** (TAAB), N,N,N',N'-tetraallylpiperazinium dibromide **1b** (TAPB) and N,N,N',N'-tetraallyltrimethylenedipiperidine dibromide **1c** (TAMPB) were synthesised by using a previously published procedure [12]. Briefly, TAAB was prepared by reacting triallylamine with allyl bromide in acetone (80 °C, 72 h). The synthesis of **1b** was performed in two steps. First, piperazine was reacted with 2 eq. allyl bromide in H_2O (20 °C, 48 h). The purification resulted in a yellow oily liquid. For the second step, the liquid was heated with 2.2 eq. allyl bromide in acetone (80 °C, 72 h) to give TAPB. Finally, the preparation of **1c** follows the two-step procedure outlined for TAPB. Trimethylenedipiperidine with 2 eq. allyl bromide afforded crude diallyltrimethylenedipiperidine (DAMP) as a brown oily liquid. In the second step, DAMP yielded TAMPB as a beige crystalline solid after recrystallization in methanol.

4.3. NMR Spectroscopy

^{1}H-NMR spectra were recorded with 400 MHz on a Mercury 400 spectrometer (Varian, Palo Alto, CA, USA). Chemical shifts were calculated using the HDO signal at 4.64 ppm or the $CDCl_3$ signal at 7.26 ppm as a reference. To check for potential alkaline hydrolysis, 25 mg of the crosslinkers were added to 1 mL of a solution containing 561.1 mg (0.01 mol) KOH in 10 mL D_2O. ^{1}H-NMR spectra were taken on the same day and after 24 h, 48 h, 72 h, and after 7 days. Spectra recorded in pure D_2O were used as reference.

4.4. Preparation of Diallyldimethylammonium Hydroxide (DADMAOH) by Ion Exchange [8]

A total of 1125 g of the ion-exchange resin in the chloride form and 1.8 L of 1 M NaOH were placed in a chromatography column. The mixture was allowed to sit for 30 min before the column was drained and the resin washed with 1.2 L bidistilled water. For the exchange, 360 g of the 65 wt % commercial diallyldimethylammonium chloride (DADMAC) solution were diluted with 750 mL water and then slowly fed onto the column followed by 2.25 L bidistilled water. The combined solutions were adjusted to a concentration of 30 wt % by rotary evaporation. The dry weight was determined gravimetrically by freeze-drying 5 mL of the solution. The amount of chloride left in the product was determined by potentiometric titration and was for all samples less than 5 mol%.

4.5. Copolymerisation of Crosslinked DADMAOH Hydrogels

The method is described using 2 mol% of the TAAB (sample TAAB2) crosslinker as an example. Details on all compositions can be found in Table S1 in the Supporting Information. A mixture of 5 g of a DADMAOH solution (30 wt % in water, 10.5 mmol), 30 mg sodium disulfite (0.16 mmol), and 54 mg tetraallylammonium bromide (0.2 mmol, 2 mol% relative to the DADMAOH content) was stirred until the crosslinker had completely dissolved. Meanwhile, 60 mg KPS was dissolved separately in 1.5 mL H_2O and then added to the monomer solution. All solutions were stirred for 15 min and then stored at room temperature for 3 weeks.

4.6. Swelling Experiments (Teabag Tests)

All tested gels were cut into small particles and lyophilized under reduced pressure. Approx. 100 mg of the dried hydrogels were weighed into polyester filter-bags and submerged in 0.5 L of bidistilled water at 22 °C. Every hour, the teabags were removed from the solution and carefully stripped off. To drain unabsorbed water, the samples were hung up for 5 min before the mass changes were measured gravimetrically. The reported maximum swelling ratios were determined by averaging the last three recorded values.

4.7. Preparation of Crosslinked DADMAOH-Co-Methacrylamide Hydrogels and Durability Test

The method is described using 2 mol% TAAB (sample TAAB2 + MAA) as an example. Details on all used compositions can be found in Table S2 in the Supporting Information. A mixture of 5 g of a DADMAOH solution (30 wt % in water, 10.5 mmol), 30 mg sodium disulfite (0.16 mmol), 54 mg tetraallylammonium bromide (0.2 mmol, 2 mol% related to the DADMAOH content), and 71.3 mg methacrylamide (0.8 mmol, 8 mol%) was stirred until the crosslinker had completely dissolved. Meanwhile, 60 mg KPS was dissolved separately in 1.5 mL H_2O and then added to the monomer solution. The solutions were stirred for 15 min and then sealed with parafilm to avoid drying effects. After storing the samples at room temperature for 3 weeks, half of the samples were placed in a Memmert UN55 drying oven at 60 °C and kept there over a period of 4 weeks. The samples were evaluated by rheological experiments according to Section 4.8 or by swelling experiments according to Section 4.6 after lyophilisation. For that, the glass vial of a sample was broken, and the gel was carefully removed. A 3–5 mm thick gel slice was then cut from the centre of the gel cylinder. Care was taken to ensure that the thickness of the slice remained the same over the entire surface. The first (reference) measurement was made after the initial 3 weeks at RT. Further measurements were made after an additional 6 weeks at RT, as well as after 7/14/21 and 28 days of storage at 60 °C. At the corresponding time points, the sample vials were carefully shattered, and the bulk gel was removed from the glass fragments.

4.8. Rheology

The rheological data were recorded on an Anton Paar Modular Compact Rheometer 102. A plate–plate geometry with a diameter of 25 mm and made of stainless steel was used. All samples were measured at 20 °C. The gap distance was selected depending on the applied normal force, which was set between 0 and 5 N depending on the experiment. For each measurement, an amplitude sweep was performed, with an angular frequency of 1 Hz and an increasing amplitude of 0.01–100% over 25 measurement points.

4.9. Ratio Exchange Tests of Crosslinked DADMAOH-Co-Methacrylamide Hydrogels

Three separate solutions were prepared from the 30% DADMAOH solution obtained in Section 4.4, in which equimolar 0/8/20 mol% DADMAOH monomers were exchanged by dilution with bidistilled water and the subsequent addition of methacrylamide. For each of the three solutions, 3 reaction mixtures for each of the crosslinkers were prepared as follows. For each sample, a mixture of 5 g of one of the above three DADMAOH-MAA-solutions (10.5 mmol), 30 mg sodium disulfite (0.16 mmol), 270 mg TAAB, or 557.5 mg TAMPB (0.2 mmol, 10 mol% related to the DADMAOH-MAA content), was stirred until the crosslinker had completely dissolved. Meanwhile, 60 mg KPS was dissolved separately in 1.5 mL H_2O and then added to the monomer solution. All solutions were stirred for 15 min and then sealed with parafilm to avoid drying effects. All samples were stored at room temperature for 3 weeks and then analyzed for storage moduli and swelling ratios according to the above procedures.

Supplementary Materials: The following are available online at https://www.mdpi.com/article/10.3390/gels8100669/s1, Figure S1: ^{1}H-NMR spectrum in $CDCl_3$ of the supernatant was performed during the decomposition of **1b** in 1 M KOH; Figure S2: Storage moduli of poly(DADMAOH-*co*-MAA) hydrogels crosslinked with 2 mol% TAAB, 2 mol% TAMPB, and 4 mol% BIS at different times after storage of the polymerized gels at room temperature; Figure S3: Swelling ratios of poly(DADMAOH-*co*-MAA)-hydrogels crosslinked with 2%$_{mol}$ TAAB/2%$_{mol}$ TAMPB/4%$_{mol}$ BIS at different times after storage of the polymerized gels at room temperature; Table S1: Details on the sample composition used to prepare crosslinked DADMAOH hydrogels; Table S2: Details on the sample composition used to prepare crosslinked DADMAOH-*co*-MAA hydrogels.

Author Contributions: All authors contributed to the study conception and design. Material preparation, data collection, and analysis were performed by T.B.M.; Original draft preparation, T.B.M.; Review and editing, O.W. All authors have read and agreed to the published version of the manuscript.

Funding: This research was funded by the Federal Ministry of Education and Research (BMBF) under grant No. 13XP5131D.

Institutional Review Board Statement: Not applicable.

Informed Consent Statement: Not applicable.

Data Availability Statement: Data are available upon reasonable request.

Acknowledgments: The authors thank Lena Schmitz and Kaja Kensmann for technical assistance.

Conflicts of Interest: The authors declare no conflict of interest.

References

1. Ahmed, E.M. Hydrogel: Preparation, characterization, and applications: A review. *J. Adv. Res.* **2015**, *6*, 105–121. [CrossRef] [PubMed]
2. Yoshioka, H.; Mori, Y.; Shimizu, M. Separation and recovery of DNA fragments by electrophoresis through a thermoreversible hydrogel composed of poly(ethylene oxide) and poly(propylene oxide). *Anal. Biochem.* **2003**, *323*, 218–223. [CrossRef] [PubMed]
3. El-Sherbiny, I.M.; Yacoub, M.H. Hydrogel scaffolds for tissue engineering: Progress and challenges. *Glob. Cardiol. Sci. Pract.* **2013**, *2013*, 316–342. [CrossRef] [PubMed]
4. Buchanan, K.J.; Hird, B.; Letcher, T.M. Crosslinked poly(sodium acrylate) hydrogels. *Polym. Bull.* **1986**, *15*, 325–332. [CrossRef]
5. Campbell, R.L.; Seymour, J.L.; Stone, L.C.; Milligan, M.C. Clinical studies with disposable diapers containing absorbent gelling materials: Evaluation of effects on infant skin condition. *J. Am. Acad. Dermatol.* **1987**, *17*, 978–987. [CrossRef]
6. Wong, H.S. Concrete with superabsorbent polymer. In *Eco-Efficient Repair and Rehabilitation of Concrete Infrastructures*; Woodhead Publishing: Cambridge, UK, 2018; pp. 467–499.
7. Akashi, M.; Saihata, S.; Yashima, E.; Sugita, S.; Marumo, K. Novel nonionic and cationic hydrogels prepared from N-vinylacetamide. *J. Polym. Sci. Part A Polym. Chem.* **1993**, *31*, 1153–1160. [CrossRef]
8. Jung, A.; Weichold, O. Preparation and characterisation of highly alkaline hydrogels for the re-alkalisation of carbonated cementitious materials. *Soft Matter* **2018**, *14*, 8105–8111. [CrossRef] [PubMed]
9. Jung, A.; Faulhaber, A.; Weichold, O. Alkaline hydrogels as ion-conducting coupling material for electrochemical chloride extraction. *Mater. Corros. Werkst. Korros.* **2021**, *72*, 1448–1455. [CrossRef]
10. Jung, A.; Weichold, O. A 3-in-1 alkaline gel for the crack injection in cement-based materials with simultaneous corrosion protection and re-passivation of crack-crossing steel rebars. *Constr Build. Mater.* **2022**, *344*, 128092. [CrossRef]
11. MacDonald, R.J. Anion Selective Polymers Prepared from Concentrated Solutions of N,N'-methylenebisacrylamide. U.S. Patent 5037858A, 6 August 1991.
12. Mrohs, T.B.; Weichold, O. Multivalent allylammonium-based cross-linkers for the synthesis of homogeneous, highly swelling diallyldimethylammonium chloride hydrogels. *Gels* **2022**, *8*, 100. [CrossRef] [PubMed]
13. Yin, Y.-L.; Prud'homme, R.K.; Stanley, F. Relationship between poly(acrylic acid) gel structure and synthesis. In *Polyelectrolyte Gels*; ACS Symposium Series; American Chemical Society Publication: Washington, DC, USA, 1992; pp. 91–113.
14. Biçeak, N.; Koza, G. A nonhydrolyzable-water soluble crosslinker: Tetraallylpiperazinium dichloride and its copolymers with acrylic acid and acrylamide. *J. Macromol. Sci. A* **1996**, *33*, 375–380. [CrossRef]
15. Zohuriaan-Mehr, M.J.; Kabiri, K. Superabsorbent polymer materials: A review. *Iran. Polym J.* **2008**, *17*, 451–477.
16. Jung, A.; Endres, M.B.; Weichold, O. Influence of environmental factors on the swelling capacities of superabsorbent polymers used in concrete. *Polymers* **2020**, *12*, 2185. [CrossRef] [PubMed]
17. Meyvis, T.K.L.; De Smedt, S.C.; Demeester, J.; Hennink, W.E. Rheological monitoring of long-term degrading polymer hydrogels. *J. Rheol.* **1999**, *43*, 933–950. [CrossRef]
18. Karpushkin, E.; Dušková-Smrčková, M.; Remmler, T.; Lapčíková, M.; Dušek, K. Rheological properties of homogeneous and heterogeneous poly(2-hydroxyethyl methacrylate) hydrogels. *Polym. Int.* **2012**, *61*, 328–336. [CrossRef]

Article

Smart Antifreeze Hydrogels with Abundant Hydrogen Bonding for Conductive Flexible Sensors

Bailin Dai [†], Ting Cui [†], Yue Xu, Shaoji Wu, Youwei Li, Wu Wang, Sihua Liu, Jianxin Tang * and Li Tang *

National & Local Joint Engineering Research Center for Advanced Packaging Material and Technology, College of Life Sciences and Chemistry, Hunan University of Technology, Zhuzhou 412007, China; daibailin7@163.com (B.D.); cuiting1132@163.com (T.C.); moonxu777@163.com (Y.X.); wushaoji321@163.com (S.W.); 13574763350@163.com (Y.L.); cvery@hut.edu.cn (W.W.); sihua0526@hut.edu.cn (S.L.)
* Correspondence: jxtang0733@163.com (J.T.); tangli_352@163.com (L.T.)
† These authors contributed equally to this work.

Abstract: Recently, flexible sensors based on conductive hydrogels have been widely used in human health monitoring, human movement detection and soft robotics due to their excellent flexibility, high water content, good biocompatibility. However, traditional conductive hydrogels tend to freeze and lose their flexibility at low temperature, which greatly limits their application in a low temperature environment. Herein, according to the mechanism that multi–hydrogen bonds can inhibit ice crystal formation by forming hydrogen bonds with water molecules, we used butanediol (BD) and N–hydroxyethyl acrylamide (HEAA) monomer with a multi–hydrogen bond structure to construct LiCl/p(HEAA–co–BD) conductive hydrogel with antifreeze property. The results indicated that the prepared LiCl/p(HEAA–co–BD) conductive hydrogel showed excellent antifreeze property with a low freeze point of −85.6 °C. Therefore, even at −40 °C, the hydrogel can still stretch up to 400% with a tensile stress of ~450 KPa. Moreover, the hydrogel exhibited repeatable adhesion property (~30 KPa), which was attributed to the existence of multiple hydrogen bonds. Furthermore, a simple flexible sensor was fabricated by using LiCl/p(HEAA–co–BD) conductive hydrogel to detect compression and stretching responses. The sensor had excellent sensitivity and could monitor human body movement.

Keywords: intelligent gel; antifreeze conductive hydrogel; flexible sensor

Citation: Dai, B.; Cui, T.; Xu, Y.; Wu, S.; Li, Y.; Wang, W.; Liu, S.; Tang, J.; Tang, L. Smart Antifreeze Hydrogels with Abundant Hydrogen Bonding for Conductive Flexible Sensors. *Gels* **2022**, *8*, 374. https://doi.org/10.3390/gels8060374

Academic Editor: Hyun-Joong Chung

Received: 15 May 2022
Accepted: 10 June 2022
Published: 13 June 2022

Publisher's Note: MDPI stays neutral with regard to jurisdictional claims in published maps and institutional affiliations.

Copyright: © 2022 by the authors. Licensee MDPI, Basel, Switzerland. This article is an open access article distributed under the terms and conditions of the Creative Commons Attribution (CC BY) license (https://creativecommons.org/licenses/by/4.0/).

1. Introduction

Hydrogel is a kind of soft−wet material with a three−dimensional polymer network [1–3]. In recent years, hydrogels have been widely applied in the fields of intelligent sensing [4,5], biomedicine [6–10], and human health monitoring [11–13] due to its high stretchability, ionic conductive path, and high water content [14–16]. Electronic or ionic conductivity is a necessary condition to ensure that hydrogel can transform external mechanical stimulation into easily collected electrical signals. However, most hydrogels lack conductivity due to the insulating properties of conventional polymer networks. Commonly used methods to enhance the conductivity of hydrogel include filling conductive fillers (graphene [17,18], carbon nanotubes [19,20], polypyrrole [21,22] and polyaniline [23], etc.) or adding soluble electrolytes (e.g., LiCl, NaCl, and $(NH_4)_2SO_4$) [24–28]. Although the introduction of conductive fillers provides ionic or electronic conductivity for hydrogels, it also raises some issues, such as difficult dispersion of conductive nanomaterials and easy agglomeration, leading to the loss of mechanical properties and electrical conductivity of hydrogels. In addition, the inherent incompatibility between a rigid conductive polymer network and flexible networks results in a decrease in the flexibility of hydrogels. On the other hand, the high water content of hydrogel makes it easy to freeze at low temperature, losing its flexibility and conductivity. Therefore, how to endow hydrogels with conductivity and improve their frost resistance remains a huge challenge.

The introduction of a soluble electrolyte (e.g., LiCl, NaCl, and $(NH_4)_2SO_4$) is one of the most commonly used methods, which can not only improve the conductive ability of hydrogels, but also inhibit the generation of ice crystals, and reduce the freezing point of hydrogels [29–31]. Zhang and co-workers prepared anti–freezing hydrogels by soaking hydrogel in $(NH_4)_2SO_4$ solution to replace the water of the hydrogel. The resultant hydrogel showed excellent ionic conductivity (~2.7 S/m) at −40 °C, and maintained good flexibility in a wide temperature range (−40~25 °C) [32]. Nevertheless, too much salt in a hydrogel will lead to undesirable mechanical properties, which is not conducive to practical applications of hydrogels. Another way to endow hydrogel with frost resistance is to introduce an organic solvent, which reduces the freezing point of hydrogels through the interaction of organic solvents with water molecules. Based on this strategy, Wu et al. designed a gelatin/N−hydroxyethyl acrylamide/glycerol/lithium chloride double network hydrogel by one pot−heating−cooling method [33]. The prepared hydrogel possesses excellent mechanical property (tensile stress/strain of 2.14 MPa/1637.49%). Moreover, due to the introduction of glycerol, the hydrogel shows good anti−freezing ability. It still has good flexibility at −40 °C and can light LED at −80 °C. However, the introduction of an organic solvent with a low dielectric coefficient reduces the ionic moving speed of a hydrogel, resulting in low conductivity the hydrogel in question.

Herein, we designed a LiCl/p(HEAA−co−BD) conductive hydrogel with anti−freezing ability and a high mechanical property. BD is a monomer with polyol structure, which is similar to ethylene glycol and glycerol. The polyol structure of BD is easy to combine with free water through hydrogen bonding, thus inhibiting the formation of ice crystals and reducing the freezing temperature of hydrogel. Moreover, HEAA was selected due to its rich amide and hydroxyl groups, which can form more hydrogen bonds between inter− and intra− polymer chains, thus enhancing the mechanical properties and adhesion properties of hydrogel. Therefore, the obtained LiCl/p(HEAA−co−BD) conductive hydrogel has a low freeze point at −85.6 °C, and the tensile stress/strain of the hydrogel can reach 450 KPa/400% at −40 °C. In addition, hydrogel shows excellent adhesive ability (~150 KPa) on various non−porous substrates, such as rubber, plastic, foam, paper and other substrates. Furthermore, a flexible sensor based on a LiCl/p(HEAA−co−BD) conductive hydrogel was prepared, which can effectively respond to different pressure and monitor human limb bending. This design strategy of a low−temperature conductive hydrogel will effectively expand the application of hydrogel.

2. Results and Discussion

2.1. Preparation of LiCl/p(HEAA−co−BD) Antifreeze Conductive Hydrogel

As shown in Figure 1a, LiCl/p(HEAA−co−BD) antifreeze conductive hydrogels were synthesized by a photopolymerization method (Figure 1). The specific synthesis steps are as follows: BD, HEAA, MBA, LiCl and I2959 were added to a bottle with water. After that, the prepared solution was injected into the mold through a syringe and placed under 8 W UV light for 1 h. In this process, free radicals were generated by the decomposition of I2959 to induce the copolymerization of BD and HEAA, which further formed a three−dimensional polymer network by cross−linking agent MBA. Among them, LiCl is uniformly dispersed in the polymer hydrogel network, which gives the hydrogel excellent electrical conductivity by forming movable free hydration ions. In addition, the hydrogel exhibited excellent freezing resistance by inhibiting the formation of ice crystals in hydrogels through the multiple hydrogen bonding of butanediol and N−hydroxyethyl acrylamide with H_2O at low temperature. As shown in Figure 1b, a pure pHEAA hydrogel ($m_{HEAA}:m_{H2O}$ = 1:1) was frozen at −20 °C, showing poor frost resistance. The formed pure pHEAA hydrogel did not freeze at −20 °C when HEAA concentration was increased to 61.5 wt% ($m_{HEAA}:m_{H2O}$ = 8:5). Obviously, with the increase in HEAA concentration, the content of hydrophilic functional groups in the pHEAA polymer chain increased, and the number of hydrogen bonds formed with free water increased, thus enhancing the freezing resistance of the polymer.

On the side, the transparency of the LiCl/p(HEAA−co−BD) anti−freezing conductive hydrogel was measured with a UV−vis spectrometer. As shown in Figure 1c, the transmission of the hydrogel was as high as 80–90% in the wavelength range of 400–700 nm, which was close to the transmission rate of some commercial transparent films. In addition, the Chinese characters (Chinese name of Hunan University of Technology) printed on the paper were clearly observed through the cuvette containing the hydrogel, and the characters of "Hunan University of Technology" were clearly observed. Compared with the blank next to them, there was no obvious blurring of the characters, showing high permeability.

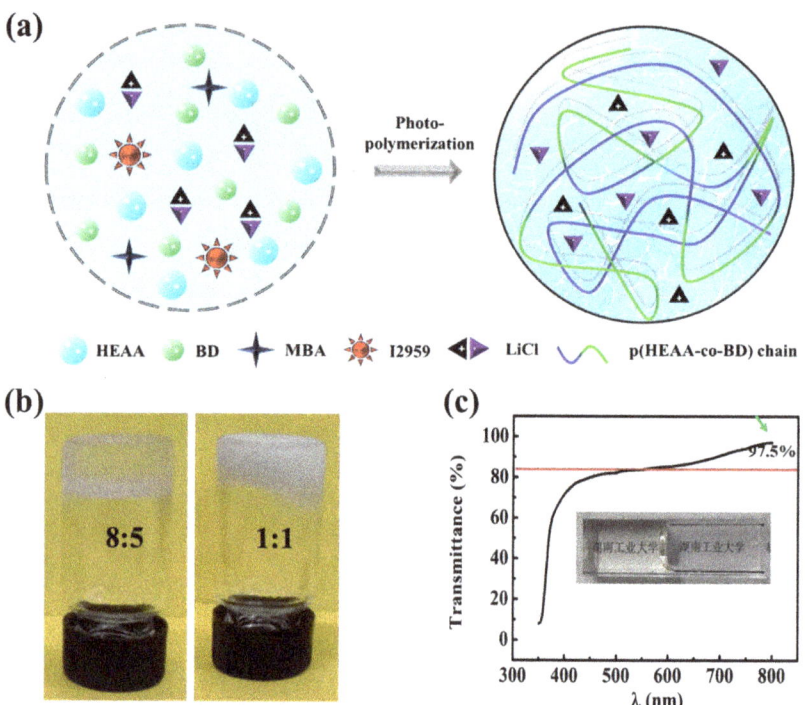

Figure 1. (a) Schematics of preparation of LiCl/p(HEAA−co−BD) antifreeze conductive hydrogel; (b) Frost resistance of HEAA hydrogels with different concentrations of HEAA at −20 °C; (c) The transmittance of LiCl/p(HEAA−co−BD) antifreeze conductive hydrogel in a colorimetric dish.

2.2. Comparison of Anti−Freezing Capacity of Different Hydrogel Components

The anti−freezing ability of different hydrogel components was shown in Table 1. All hydrogel prepolymerized solutions were transparent. Then, the prepolymerized solution was polymerized by UV light to form a transparent hydrogel at room temperature (RT). The anti−freezing ability of the hydrogel was judged by the change of its color at −20 °C. The hydrogel still showed high light transmittance after being frozen at −20 °C for 12 h, which indicated that the hydrogel did not freeze at −20 °C. When this series of hydrogels were further frozen at −80 °C for 12 h, it was found that the LiCl/p(HEAA−co−BD)/H$_2$O hydrogel was still transparent, which means that there was no freezing in this hydrogel, while all or part of the LiCl/pHEAA/H$_2$O, p(HEAA−co−BD)/H$_2$O and pHEAA/H$_2$O hydrogels were milky white. This means that the LiCl/p(HEAA−co−BD)/H$_2$O hydrogel had the best anti−freezing ability, which was attributed to the interaction between the monomer and LiCl with water molecules.

Table 1. Freezing resistance of each component hydrogel.

	LiCl/p(HEAA−co−BD)/H₂O	LiCl/pHEAA/H₂O	LiCl/pHEAA/H₂O	pHEAA/H₂O
Monomeric ratios (g:g)	0.45:8:2:5	0.45:8:5	8:2:5	8:5
Pre−polymerized liquid				
Post−aggregation				
−20 °C				
−80 °C				
Glass transition temperature	−85.6 °C	−51.8 °C	−64.9 °C	−17.1 °C

2.3. Low−Temperature Tensile Properties of LiCl/p(HEAA−co−BD) Anti−Freezing Conductive Hydrogels

Tensile properties of the LiCl/p(HEAA−co−BD) hydrogel at different temperatures were shown in Figure 2a. At RT, −20 °C and −80 °C, the hydrogel shows good tensile properties and excellent flexibility.

In order to quantitatively test and compare the anti−freezing ability of this series of hydrogels, we presented the DSC curves of hydrogels (Figure 2b). The glass transition temperatures of the LiCl/pHEAA/H₂O, p(HEAA−co−BD)/H₂O and pHEAA/H₂O hydrogels were −51.8 °C, −64.9 °C and −17.1 °C, respectively. The above data demonstrated that both BD and LiCl have excellent anti−freezing ability. It can be clearly observed that after combining BD and LiCl with pHEAA/H₂O hydrogel, the resultant LiCl/p(HEAA−co−BD)/H₂O hydrogel had the lowest glass transition temperature (Tg) of −85.6 °C. This method overcomes the limitation that traditional antifreeze hydrogels are mostly based on organic hydrogels and obtains excellent antifreeze water−based hydrogels.

The tensile properties of the hydrogel were quantitatively characterized by a tensile test at a low temperature from RT to −40 °C (Figure 2c). The hydrogel showed superb tensile properties, which can be stretched up to 200 KPa/1000% at RT. When the temperature was gradually reduced to −10 °C, −20 °C and −30 °C, the fracture strain of the hydrogel did not change significantly, but the fracture stress gradually increased. When the temperature was reduced to −40 °C, the fracture strain shrank to ~400% and the fracture stress reached a maximum of 450 KPa. Obviously, when the temperature decreases from RT to −40 °C, the elastic modulus of the hydrogel showed a similar trend to the tensile properties of the hydrogel (Figure 2d). This occurs as, when the temperature drops, the hydrogel gradually begins to change from a highly elastic state to a glassy state, where it becomes harder and less flexible.

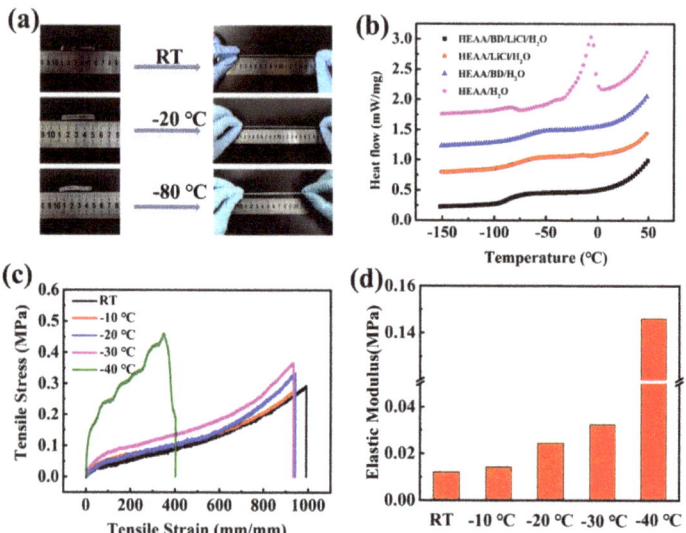

Figure 2. (a) LiCl/p(HEAA−co−BD) hydrogel stretched under RT, −20 °C and −80 °C; (b) DSC curves of hydrogels with different components; (c) Stress–strain curves and; (d) Elastic modulus of LiCl/p(HEAA−co−BD) hydrogel at RT, −10 °C, −20 °C, −30 °C, and −40 °C, respectively.

2.4. Adhesion of LiCl/p(HEAA−co−BD) Hydrogel

The LiCl/p(HEAA−co−BD) hydrogel possesses excellent adhesion due to it contains a large number of hydrogen bonds. First, the adhesive properties of hydrogel on various non−porous substrates were qualitatively tested. As shown in Figure 3a, the LiCl/p(HEAA−co−BD) hydrogel not only could adhere to inorganic materials such as stainless steel, ceramics and glass, but also can adhere to organic materials such as rubber, plastic, foam and cardboard. In addition to showing adhesion to hydrophilic substances, the hydrogel also adhered to hydrophobic PTFE substrates, overcoming the problem that most adhesives do not adhere to hydrophobic materials. Furthermore, the adhesion strength of the hydrogel on glass substrates was quantitatively tested. Figure 3b shown a schematic diagram of 180° peeling test for exploring the adhesion of hydrogel [34]. The adhesion strength of the LiCl/p(HEAA−co−BD) hydrogel at different peeling speeds is shown in Figure 3c. It can be seen that the adhesion strength of the hydrogel gradually increased from ~60 KPa to 150 KPa as the peeling speed increased from 10 mm/min to 100 mm/min, showing obvious speed−dependent properties.

More than that, the LiCl/p(HEAA−co−BD) hydrogel demonstrated excellent reversible adhesion properties. As shown in Figure 3d, the reversible adhesive strength of the hydrogel was tested five times with a resting time of two minutes during the two peeling tests. The results showed that the hydrogel adhesion strength decreased rapidly from ~160 KPa to ~50 KPa when the second adhesion test was performed. This occurs as, in the early stage of synthesis, the prepolymerized solution could be filled into the micropores on the glass substrate, which can eliminate some invisible bubbles on the glass substrate. After an hour photopolymerization, the hydrogel fragments in the micropores form a whole with the surface hydrogel, so the strength of hydrogel is higher than cyclic adhesion. After the first peeling, some hydrogel fragments would remain on the surface of the hydrogel, and when re−adhering, these hydrogel fragments would contact with the bulk hydrogel to form bubbles, thus reducing the adhesion ability when re−adhering. Due to the above reasons, the second adhesive strength will be lower than the first adhesive strength. When the peeling−adhesion test was repeated several times, the adhesion strength was basically stabilized at ~30 KPa. This is due to the immediate recovery of dynamic reversible interaction. Therefore, the LiCl/p(HEAA−co−BD) hydrogel showed relatively weak and relatively stable adhesive strength in a limited recovery time.

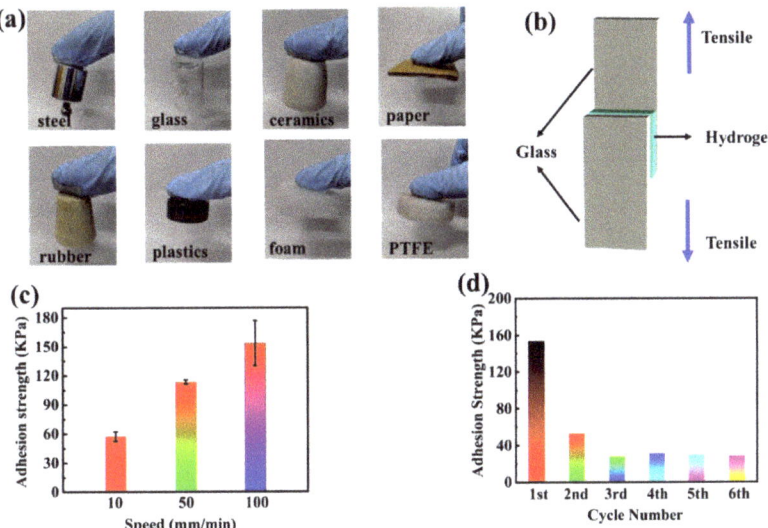

Figure 3. (**a**) LiCl/p(HEAA−co−BD) hydrogels adhere to different substrates; (**b**) Schematic diagram of lap shear testing for strip strength; (**c**) Effect of peel speed on adhesion strength of hydrogels; (**d**) multiple adhesion of hydrogels.

2.5. Conductivity of LiCl/p(HEAA−co−BD) Antifreeze Hydrogels

The conductivity of the LiCl/p(HEAA−co−BD) hydrogel was tested by serially connecting it into the circuit of a battery with 8V at RT, −20 °C and −80 °C, respectively. The results were shown in Figure 4. The hydrogel not only had excellent conductivity at RT (LED emitted bright light), but also possessed conductivity at a low temperature (LED can be lit). Among them, since LiCl can be hydrogenated to freely mobile Li^+ and Cl^- at RT, which can allow the hydrogel to exhibit conductive ability, the LED in the path exhibits bright emission. With the decrease in temperature, the decline in mobility of hydrated Li^+, resulting in the decrease in the conductivity of hydrogel.

2.6. Sensing Performance of LiCl/p(HEAA−co−BD) Hydrogel

Figure 5a shows the resistance changes of LiCl/p(HEAA−co−BD) hydrogel with compressive strain. It could be seen that the resistance changes of hydrogel showed an obvious linear relationship (R^2 = 0.99) and GF = 0.65 with the compressive strain of 0–50%. Furthermore, we tested the pressure response properties of the hydrogel by adding heavy weights. As shown in Figure 5b. As the mass of the weight applied to the hydrogel increases from zero to 100 g, the resistance changes rate of hydrogel decreased from 100% to 4%, showing excellent pressure response properties.

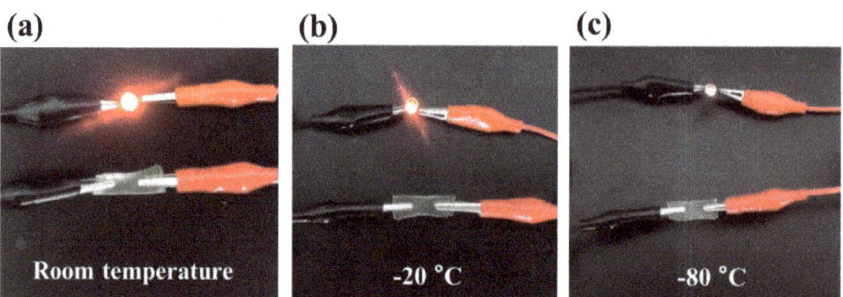

Figure 4. Conductivity of LiCl/p(HEAA−co−BD) hydrogel at different temperatures: (a) RT; (b) −20 °C and; (c) −80 °C.

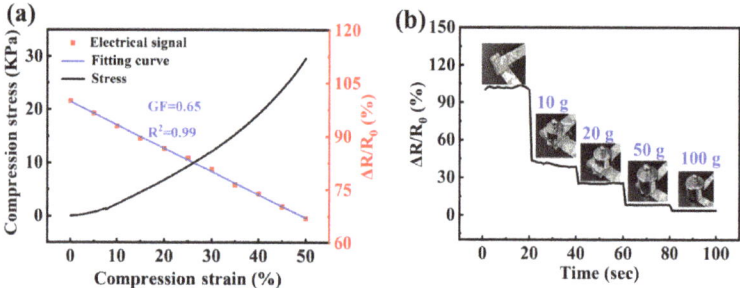

Figure 5. (a) Compression stress and compression strain and (b) Resistance changes of hydrogel response to different weights.

The sensitivity of the strain sensor, based on the LiCl/p(HEAA−co−BD) hydrogel, was also assessed. The GF were 1.12, 1.68 and 2.19 in the tensile strain range of 0–100%, 100–200% and 200–500% (Figure 6a), respectively, which exhibited excellent sensitivity. Particularly, the obvious signals were successfully gained in the process of loading–unloading under strain of 1%, 100–400%, respectively. (Figure 6b,c). Additionally, the response time of the hydrogel flexible sensor was simultaneously recorded at the stretching speed of 100 mm/min and 200% strain (Figure 6d). The results showed that the response time was only 0.2 s, and it is proved that the hydrogel flexible sensor could immediately respond to the changes in external strain. The stability of the flexible sensor was tested via sixty times loading–unloading cycles with 50% strain. As shown in Figure 6e, the relative resistance of sensor has remained almost unchanged during the cyclic deformations (Figure 6e inset). The deviation of relative resistance was less than 1%, which means that the flexible sensor based on the LiCl/p(HEAA−co−BD) hydrogel shows high stability.

In view of the high tensile property, good electrical conductivity, adhesive property and strain response capability of this anti−freezing hydrogel. A flexible wearable sensor for monitoring human motion was prepared based on the LiCl/p(HEAA−co−BD) hydrogel. As shown in Figure 7, the hydrogel based flexible sensor was directly adhered to each joint of the human body. Then, the human movement signals were collected by an electrochemical workstation. The results showed that the relative resistance of hydrogel varied from 0 to 60% with the stretching and bending of a finger, and when the wrist, elbow, and knee were straight and bent, the relative resistance changes of the hydrogel were 30%, 100%, and 50%, respectively. Since the relative deformability of fingers and elbows is greater than that of wrists and knees, the relative resistance of hydrogels changes more obviously with bending deformation.

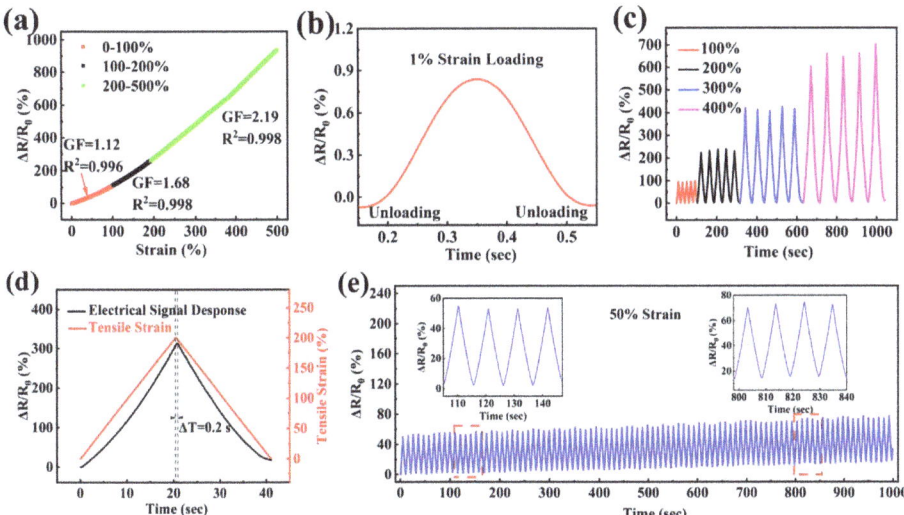

Figure 6. (a) GF of the LiCl/p(HEAA−co−BD) hydrogel flexible sensor. The change of relative resistance of the hydrogel flexible sensor during the loading–unloading cycle with a strain of (b) 1%, (c) 100–400%. (d) time−resolved responses of hydrogel flexible sensor. (e) The stability of the hydrogel flexible sensor.

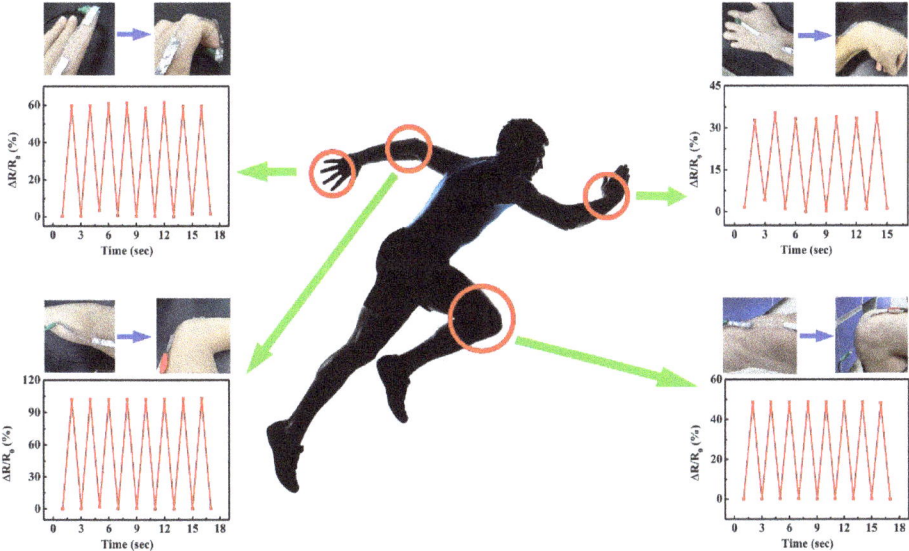

Figure 7. Applications of the LiCl/p(HEAA−co−BD) hydrogel for monitoring human motion.

3. Conclusions

In conclusion, LiCl/p(HEAA−co−BD) hydrogels with freeze resistance and electrical conductivity were prepared by one−pot photopolymerization with different antifreeze components, such as LiCl, BD and HEAA. Due to the interaction between the multi−hydrogen bonding molecular network with water molecules and the hydration of LiCl, the hydrogel showed a low freezing point of −85.8 °C. Moreover, the hydrogel exhibited excellent tensile strength (the tensile stress/strain of the hydrogel could reach 450 KPa/400% at

−40 °C) and conductivity (the LED could still be lit at −80 °C) at low temperature. In addition, based on the dynamic reversible property of a hydrogen bond, the hydrogel also showed a reversible adhesion strength of ~30 KPa. Finally, the flexible sensor based on a LiCl/p(HEAA−co−BD) hydrogel was prepared, which showed a high sensitivity (GF = 2.19) and a fast response speed (0.2 s) to external force. Based on the above advantages, the hydrogel can effectively monitor the movements of human limbs (such as finger bending, knee bending, wrist bending and elbow bending).

4. Experimental Parts

Experimental reagents and specifications. N−hydroxyethyl acrylamide (HEAA, 98%, chemically pure) was purchased from TCL Reagents Inc. 2−Hydroxy−4′−(2−hydroxyethoxy)−2−methylpropiophenone (I2959, 99%, chemically pure) and N,N′−methylene bisacrylamide (MBA, 99%, chemically pure) were purchased from Shanghai Aladdin Reagent Co. Butenediol (BD, 98%, chemically pure) was purchased from Jiangsu Aikang Biopharmaceutical R&D Co. Lithium chloride (LiCl, analytical purity) was purchased from Sinopharm Reagent Co. Deionized water was prepared by Milli−Q system water purification system, with an impedance of 18.2 MΩ/cm. All reagents were not further purified before use.

Preparation of LiCl/p(HEAA−co−BD) antifreeze conductive hydrogel. LiCl/P(HEAA−co−BD) conductive hydrogels with anti−freezing ability was prepared by photo−initiated radical polymerization reaction. The specific synthesis steps are as follows: First of all, the reactants of LiCl (0.21 g, 3 wt.%), HEAA (3.5 g, 50 wt.%), BD (0.7 g, 10 wt.%), MBA (6.8 mg), and photo−initiator I2959 (68 mg) were added to a glass bottle containing deionized water (2.8 g, 40 wt.%). Then, the mixture solution was stirred to form a clear solution. The solution was injected into a glass mold with a thickness of 1 mm through a syringe, and then transferred to an 8 W 365 nm UV light for polymerization for 1 h. After polymerization, the transparent hydrogel was formed. In order to facilitate the removal of the prepared hydrogel from the mold, a layer of PET film was covered on the contact surface between the glass and the hydrogel.

Mechanical property test. Tensile tests were carried out on a universal tester (AGS−X) equipped with 1000 N load cell at a speed of 100 mm/min. The hydrogel specimen was cut into a dumbbell shape with an effective stretching length of 25 mm, a width of 4 mm and a thickness of 1 mm, and mounted on the chuck of the stretching machine. Stretched at a rate of 100 mm/min until the hydrogel fractured. The tensile strain (ε) was calculated by the following equation:

$$\varepsilon = l_t/l_0 \tag{1}$$

where l_0 is the original length of the hydrogel and l_t is the length of the hydrogel after stretching. The tensile stress (σ) is the tensile force applied per unit area, and the formula is as follows:

$$\sigma = F/A_0 \tag{2}$$

where F is the tensile force and A_0 is the cross−sectional area of the hydrogel sample.

Low temperature tensile test. The tensile stress and tensile strain of hydrogels below zero temperature were tested by a universal tester (Zwick Roell Z010) with an environmental chamber. Before stretching, each sample was kept at the test temperature for 20 min to ensure that the temperature of the sample was the same as that of the external environment. The stretching rate was also controlled at 100 mm/min.

Differential Scanning Calorimetry (DSC). DSC test was performed using NETZSCH DSC 200F3 at a heating rate of 10 °C/min in a set temperature range under nitrogen protection.

Adhesion strength test. The adhesive strength of the hydrogels was determined using a 180° peeling test at a peeling−set speed. The hydrogels were prepared into cuboids (length, 10 mm; width, 10 mm; thickness, 2 mm). Adhesive strength (g) was calculated as follows:

$$g = F_{max}/S \tag{3}$$

where F_{max} is the maximum force applied in the stripping process, and S is the area of the hydrogel sample.

Transmittance test. The UV−vis absorption spectrum of the hydrogel was acquired using a UV−vis spectrophotometer (TU−1810) at a wavelength scanning range from 400 to 800 nm and a scanning rate of 100 nm/min.

Sensing performance test. Compression test electrochemical performance test: Compression response of conductive gels to different masses of weights: The LiCl/p(HEAA−co−BD) conductive hydrogel was made into a cylinder with a diameter of 1 cm and a height of 1 cm, and each of the two cross sections was connected to the electrochemical workstation through an aluminum sheet. The gage factor (GF) is calculated according to the following formula:

$$GF = \triangle R/R_0 \times 100\% \qquad (4)$$

where R_0 is the original resistance, and $\triangle R$ is the change of resistance during stretching.

Author Contributions: Conceptualization, J.T. and L.T.; methodology, B.D. and T.C.; software, B.D., T.C. and W.W.; validation, J.T. and L.T.; investigation, Y.X., S.W. and Y.L.; resources, S.L., J.T. and L.T.; writing—original draft preparation, B.D. and T.C.; writing—review and editing, S.L., J.T. and L.T.; visualization, Y.X. and S.W.; supervision, L.T.; project administration, J.T. and L.T.; funding acquisition, J.T. and L.T. All authors have read and agreed to the published version of the manuscript.

Funding: This research was funded by the National Natural Science Foundation of China, grant number 51774128, the Scientific research project of Hunan Education Department, grant number 21C0425.

Institutional Review Board Statement: Not applicable.

Informed Consent Statement: Not applicable.

Data Availability Statement: The data presented in this study are available on request from the corresponding author.

Conflicts of Interest: The authors declare no conflict of interest.

References

1. Tang, J.; Huang, J.; Zhou, G.; Liu, S. Versatile fabrication of ordered cellular structures double network composite hydrogel and application for cadmium removal. *J. Chem. Thermodyn.* **2020**, *141*, 105918. [CrossRef]
2. Wu, Z.; Yang, X.; Wu, J. Conductive hydrogel−and organohydrogel−based stretchable sensors. *ACS Appl. Mater. Interfaces* **2021**, *13*, 2128–2144. [CrossRef] [PubMed]
3. Zhang, D.; Tang, Y.; Zhang, Y.; Yang, F.; Liu, Y.; Wang, X.; Yang, J.; Gong, X.; Zheng, J. Highly stretchable, self−adhesive, biocompatible, conductive hydrogels as fully polymeric strain sensors. *J. Mater. Chem. A* **2020**, *8*, 20474–20485. [CrossRef]
4. Yang, C.; Su, F.; Xu, Y.; Ma, Y.; Tang, L.; Zhou, N.; Liang, E.; Wang, G.; Tang, J. pH Oscillator−Driven Jellyfish−like Hydrogel Actuator with Dissipative Synergy between Deformation and Fluorescence Color Change. *ACS Macro Lett.* **2022**, *11*, 347–353. [CrossRef]
5. Tang, L.; Zhang, D.; Gong, L.; Zhang, Y.; Xie, S.; Ren, B.; Liu, Y.; Yang, F.; Zhou, G.; Chang, Y. Double−network physical cross−linking strategy to promote bulk mechanical and surface adhesive properties of hydrogels. *Macromolecules* **2019**, *52*, 9512–9525. [CrossRef]
6. Chakraborty, P.; Oved, H.; Bychenko, D.; Yao, Y.; Tang, Y.; Zilberzwige-Tal, S.; Wei, G.; Dvir, T.; Gazit, E. Nanoengineered Peptide-Based Antimicrobial Conductive Supramolecular Biomaterial for Cardiac Tissue Engineering. *Adv. Mater.* **2021**, *33*, 2008715. [CrossRef]
7. Liu, K.; Wei, S.; Song, L.; Liu, H.; Wang, T. Conductive hydrogels—A novel material: Recent advances and future perspectives. *J. Agric. Food Chem.* **2020**, *68*, 7269–7280. [CrossRef]
8. Zhao, Y.; Zhu, Z.S.; Guan, J.; Wu, S.J. Processing, mechanical properties and bio−applications of silk fibroin−based high−strength hydrogels. *Acta Biomater.* **2021**, *125*, 57–71. [CrossRef]
9. Dong, Y.; Zhuang, H.; Hao, Y.; Zhang, L.; Yang, Q.; Liu, Y.; Qi, C.; Wang, S. Poly (N−isopropyl−acrylamide)/poly (γ−glutamic acid) thermo−sensitive hydrogels loaded with superoxide dismutase for wound dressing application. *Int. J. Nanomed.* **2020**, *15*, 1939–1950. [CrossRef]
10. Rowland, M.J.; Parkins, C.C.; McAbee, J.H.; Kolb, A.K.; Hein, R.; Loh, X.J.; Watts, C.; Scherman, O.A. An adherent tissue−inspired hydrogel delivery vehicle utilised in primary human glioma models. *Biomaterials* **2018**, *179*, 199–208. [CrossRef]
11. Xu, L.; Chen, Y.; Guo, Z.; Tang, Z.; Luo, Y.; Xie, S.; Li, N.; Xu, J. Flexible Li$^+$/agar/pHEAA double−network conductive hydrogels with self−adhesive and self−repairing properties as strain sensors for human motion monitoring. *React. Funct. Polym.* **2021**, *168*, 105054. [CrossRef]

12. Darabi, M.A.; Khosrozadeh, A.; Mbeleck, R.; Liu, Y.; Chang, Q.; Jiang, J.; Cai, J.; Wang, Q.; Luo, G.; Xing, M. Skin-inspired multifunctional autonomic-intrinsic conductive self-healing hydrogels with pressure sensitivity, stretchability, and 3D printability. *Adv. Mater.* **2017**, *29*, 1700533. [CrossRef] [PubMed]
13. Lu, J.; Han, X.; Dai, L.; Li, C.; Wang, J.; Zhong, Y.; Yu, F.; Si, C. Conductive cellulose nanofibrils−reinforced hydrogels with synergetic strength, toughness, self−adhesion, flexibility and adjustable strain responsiveness. *Carbohydr. Polym.* **2020**, *250*, 117010. [CrossRef] [PubMed]
14. Wu, S.; Tang, L.; Xu, Y.; Yao, J.; Tang, G.; Dai, B.; Wang, W.; Tang, J.; Gong, L. A self−powered flexible sensing system based on a super−tough, high ionic conductivity supercapacitor and a rapid self−recovering fully physically crosslinked double network hydrogel. *J. Mater. Chem. C* **2022**, *10*, 3027–3035. [CrossRef]
15. Tang, L.; Wu, S.; Xu, Y.; Li, Y.; Dai, B.; Yang, C.; Liu, A.; Tang, J.; Gong, L. Design of a DNA-Based Double Network Hydrogel for Electronic Skin Applications. *Adv. Mater. Technol.* **2022**, *7*, 2200066. [CrossRef]
16. Tang, L.; Wu, S.; Qu, J.; Gong, L.; Tang, J. A review of conductive hydrogel used in flexible strain sensor. *Materials* **2020**, *13*, 3947. [CrossRef]
17. He, Q.; Liu, J.; Liu, X.; Li, G.; Deng, P.; Liang, J. Manganese dioxide Nanorods/electrochemically reduced graphene oxide nanocomposites modified electrodes for cost−effective and ultrasensitive detection of Amaranth. *Colloids Surf. B Biointerfaces* **2018**, *172*, 565–572. [CrossRef]
18. Alam, A.; Meng, Q.; Shi, G.; Arabi, S.; Ma, J.; Zhao, N.; Kuan, H.C. Electrically conductive, mechanically robust, pH−sensitive graphene/polymer composite hydrogels. *Compos. Sci. Technol.* **2016**, *127*, 119–126. [CrossRef]
19. Xiang, K.; Wen, X.; Hu, J.; Wang, S.; Chen, H. Rational Fabrication of Nitrogen and Sulfur Codoped Carbon Nanotubes/MoS2 for High-Performance Lithium–Sulfur Batteries. *ChemSusChem* **2019**, *12*, 3602–3614. [CrossRef]
20. Hsiao, L.-Y.; Jing, L.; Li, K.; Yang, H.; Li, Y.; Chen, P.-Y. Carbon nanotube−integrated conductive hydrogels as multifunctional robotic skin. *Carbon* **2020**, *161*, 784–793. [CrossRef]
21. Sun, M.; Wu, X.; Liu, C.; Xie, Z.; Deng, X.; Zhang, W.; Huang, Q.; Huang, B. The In Situ grown of activated Fe−N−C nanofibers derived from polypyrrole on carbon paper and its electro−catalytic activity for oxygen reduction reaction. *J. Solid State Electrochem.* **2018**, *22*, 1217–1226. [CrossRef]
22. Zhao, L.; Li, X.; Li, Y.; Wang, X.; Yang, W.; Ren, J. Polypyrrole−Doped Conductive Self−Healing Composite Hydrogels with High Toughness and Stretchability. *Biomacromolecules* **2021**, *22*, 1273–1281. [CrossRef] [PubMed]
23. Wu, C.; Shen, L.; Lu, Y.; Hu, C.; Liang, Z.; Long, L.; Ning, N.; Chen, J.; Guo, Y.; Yang, Z.; et al. Intrinsic Antibacterial and Conductive Hydrogels Based on the Distinct Bactericidal Effect of Polyaniline for Infected Chronic Wound Healing. *ACS Appl. Mater. Interfaces* **2021**, *13*, 52308–52320. [CrossRef] [PubMed]
24. Tang, L.; Gong, L.; Xu, Y.; Wu, S.; Wang, W.; Zheng, B.; Tang, Y.; Zhang, D.; Tang, J.; Zheng, J. Mechanically Strong Metal–Organic Framework Nanoparticle−Based Double Network Hydrogels for Fluorescence Imaging. *ACS Appl. Nano Mater.* **2022**, *5*, 1348–1355. [CrossRef]
25. Xu, H.; Lv, Y.; Qiu, D.; Zhou, Y.; Zeng, H.; Chu, Y. An ultra−stretchable, highly sensitive and biocompatible capacitive strain sensor from an ionic nanocomposite for on−skin monitoring. *Nanoscale* **2019**, *11*, 1570–1578. [CrossRef]
26. Kang, T.H.; Chang, H.; Choi, D.; Kim, S.; Moon, J.; Lim, J.A.; Lee, K.Y.; Yi, H. Hydrogel−templated transfer−printing of conductive nanonetworks for wearable sensors on topographic flexible substrates. *Nano Lett.* **2019**, *19*, 3684–3691. [CrossRef]
27. Peng, H.; Xin, Y.; Xu, J.; Liu, H.; Zhang, J. Ultra−stretchable hydrogels with reactive liquid metals as asymmetric force−sensors. *Mater. Horiz.* **2019**, *6*, 618–625. [CrossRef]
28. Zhang, Q.; Liu, X.; Duan, L.; Gao, G. Ultra−stretchable wearable strain sensors based on skin−inspired adhesive, tough and conductive hydrogels. *Chem. Eng. J.* **2019**, *365*, 10–19. [CrossRef]
29. Sui, X.; Guo, H.; Cai, C.; Li, Q.; Wen, C.; Zhang, X.; Wang, X.; Yang, J.; Zhang, L. Ionic conductive hydrogels with long−lasting antifreezing, water retention and self−regeneration abilities. *Chem. Eng. J.* **2021**, *419*, 129478. [CrossRef]
30. Kong, W.; Wang, C.; Jia, C.; Kuang, Y.; Pastel, G.; Chen, C.; Chen, G.; He, S.; Huang, H.; Zhang, J.J.A.M. Muscle-Inspired Highly Anisotropic, Strong, Ion-Conductive Hydrogels. *Adv. Mater.* **2018**, *30*, 1801934. [CrossRef]
31. Liu, Y.; Wang, W.; Gu, K.; Yao, J.; Shao, Z.; Chen, X. Poly (vinyl alcohol) Hydrogels with Integrated Toughness, Conductivity, and Freezing Tolerance Based on Ionic Liquid/Water Binary Solvent Systems. *ACS Appl. Mater. Interfaces* **2021**, *13*, 29008–29020. [CrossRef]
32. Sui, X.; Guo, H.; Chen, P.; Zhu, Y.; Wen, C.; Gao, Y.; Yang, J.; Zhang, X.; Zhang, L. Zwitterionic osmolyte-based hydrogels with antifreezing property, high conductivity, and stable flexibility at subzero temperature. *Adv. Funct. Mater.* **2020**, *30*, 1907986. [CrossRef]
33. Tang, L.; Wu, S.; Xu, Y.; Cui, T.; Li, Y.; Wang, W.; Gong, L.; Tang, J. High toughness fully physical cross−linked double network organohydrogels for strain sensors with anti−freezing and anti−fatigue properties. *Mater. Adv.* **2021**, *2*, 6655–6664. [CrossRef]
34. Jing, X.; Mi, H.Y.; Lin, Y.J.; Enriquez, E.; Peng, X.F.; Turng, L.S. Highly stretchable and biocompatible strain sensors based on mussel−inspired super−adhesive self-healing hydrogels for human motion monitoring. *ACS Appl. Mater. Interfaces* **2018**, *10*, 20897–20909. [CrossRef] [PubMed]

Article

Highly Efficient Adsorption of Heavy Metals and Cationic Dyes by Smart Functionalized Sodium Alginate Hydrogels

Tianzhu Shi [1,2,*], Zhengfeng Xie [2], Xinliang Mo [1], Yulong Feng [1], Tao Peng [1] and Dandan Song [1]

1. Department of Brewing Engineering, Moutai Institute, Renhuai 564500, China; xinliangmo@163.com (X.M.); fengyulong520110@163.com (Y.F.); edifcztony@126.com (T.P.); vi_veneto@163.com (D.S.)
2. Oil & Gas Field Applied Chemistry Key Laboratory of Sichuan Province, College of Chemistry and Chemical Engineering, Southwest Petroleum University, Chengdu 610500, China; xiezhf@swpu.edu.cn
* Correspondence: shitianzhu1018@163.com; Tel.: +86-185-8642-0308

Abstract: In this paper, functionalized sodium alginate hydrogel (FSAH) was prepared to efficiently adsorb heavy metals and dyes. Hydrazide-functionalized sodium alginate (SA) prepared hydrazone groups to selectively capture heavy metals (Pb^{2+}, Cd^{2+}, and Cu^{2+}), and another functional group (dopamine grafting), serves as sites for adsorption methylene blue (MB), malachite green (MG), crystal violet (CV). Thermodynamic parameters of adsorption indicated that the adsorption process is endothermic and spontaneous. The heavy metals adsorption by FSAH was physical adsorption mainly due to $\Delta H^{\theta} < 40$ kJ/mol, and the adsorption of cationic dyes fitted with the Langmuir models, which indicated that the monolayer adsorption is dominated by hydrogen bonds, electrostatic interactions, and π-π interactions. Moreover, the adsorption efficiency maintained above 70% after five adsorption-desorption cycles. To sum up, FSAH has great application prospect.

Keywords: sodium alginate; modification; heavy metals; dye; adsorption

1. Introduction

In the process of rapid industrial development, it is easy to cause environmental pollution problems. The improper treatment of industrial waste water leads to the destruction of the water ecological environment [1]. In particular, heavy metals and dyes in sewage are very resistant to biodegradation, posing a serious threat to human health [2], for example, lead poisoning can cause brain damage and kidney and liver dysfunction; excessive malachite green can also cause nausea, abdominal pain, etc. [3–7]. The adsorption method is widely used due to its advantages of low-cost, simple-operation, and good selectivity. Heavy metals and dyes coexist in wastewater, and the current adsorption material treatment capacity limits its large-scale application [8–11]. Researchers are committed to developing adsorbents with large adsorption capacity, no secondary pollution, and wide application.

SA, a natural polysaccharide, is widely used in the food industry, pharmaceutical industry, rubber industry, and other industries. At the same time, SA as adsorption material to adsorb heavy metals and organic pollutions through physical and chemical modification, has attracted wide attention [12,13]. The chemical modification of SA can be divided into: the C–C bond of o-diol that was oxidized to dialdehyde or dicarboxylic groups; and COO^- on SA was grafted to new groups, such as amides or esters [14].

Hydrazone is very sensitive to heavy metals and can easily coordinate to form coordination structures, [15–18]. DA (dopamine) has strong functional modification ability, and catechol and amine groups could form various hydrogen bonds to adsorb target pollutants [19–21]. DA-modification is a highly functional surface modification strategy. Combining the advantages of the hydrazone group and DA-modification, an SA-based adsorption material FSAH was prepared.

In this paper, the functional modification of sodium alginate was divided into two steps: the hydrazone group was prepared by using sodium alginate and biphthalate dihydrazide (BDD) based on the Schiff base reaction, and DA was grafted to form an amide carbonyl group to prepare the adsorbent FSAH. The adsorption capacity of heavy metals and dyes was evaluated, and pH, initial concentration, time, and reaction temperature were investigated. The adsorption isotherm equations, adsorption thermodynamics, and adsorption kinetics were analyzed in the adsorption process. According to FT-IR, SEM, and XPS analysis, the mechanism of FSAH adsorption of heavy metals and dyes was proposed. Adsorption–desorption experiments showed FSAH had broad application prospects in the removal of heavy metals and dyes.

2. Results and Discussion
2.1. Characterizations
2.1.1. SEM and EDS–Mapping Analysis

The microstructure and morphology of material before and after modification were recorded by SEM, respectively. The SA surface was smooth with dense small particles (Figure 1a), but the surface of FSAH (Figure 1b) showed rough, multi-fractured, and porous structure resulting from SA that was grafted with BDD and DA, respectively, and the specific surface area (BET) was increased. Figure 1c,d shows the C, O, and Na elements in SA by EDS–mapping as shown in Table S1, but the N element is added to FSAH after modification. The element content of C increased from 32.49 to 48.35%, however, O and Na contents decreased, respectively. The new element N increased to 25.88%, as shown in Table S5.

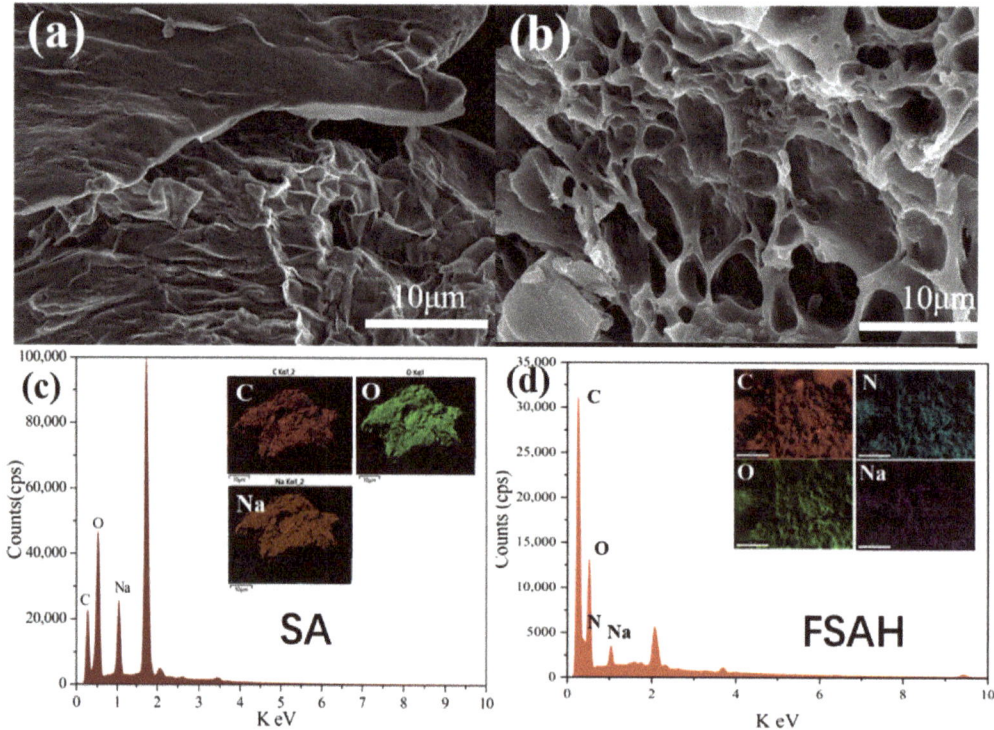

Figure 1. SEM of SA (a) and FSAH (b); EDS–mapping of SA (c), FSAH (d).

2.1.2. FT-IR and XPS Analysis

The infrared absorption spectra of FSAH, DSA (dialdehyde sodium alginate), and SA were determined using FT-IR spectroscopy (Figure 2a). The dialdehyde group (1731 cm^{-1}) was generated by oxidation of the C–C bond in o-diol groups by NaIO$_4$, and the stretching vibration of the aldehyde group proton was observed at 2910 and 2842 cm^{-1} of DSA [22]. In FSAH, the C=N imine and the N-H peak occurred at 1597 and 1003 cm^{-1}, respectively, and the 1731 cm^{-1} peak vanished, indicating that the hydrazone group was successfully prepared [23]. The connection between δ_{N-H} and ν_{C-N}, as well as the C=N peak positions, created the amide II (1528 cm^{-1}) and amide III (1283 cm^{-1}) belts peaks, respectively.

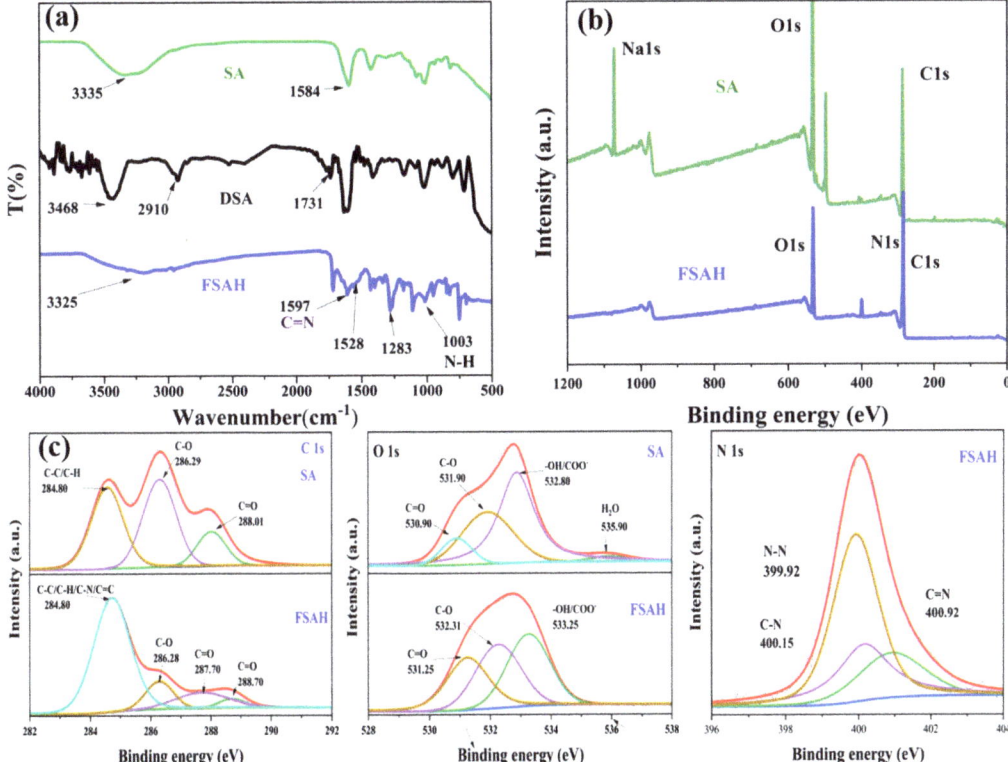

Figure 2. (a) FT–IR spectra of SA, DSA, and FSAH. (b) The XPS spectra of SA and FSAH. (c) C 1s, O 1s, and N 1s spectra of SA and FSAH.

As shown in Figure 2b,c, C 1s, O 1s, and N 1s core levels were investigated using deconvoluted fitting of complex spectra, and the N 1s, which was new peak, occurred in FSAH. The C 1s spectrum for SA displayed the bonds as being C–C, C–H bond (284.80 eV), O–C–O bond (286.29 eV), C–OH bond (286.88 eV), and C=O bond (288.01 eV), respectively [24,25]. However, a new peak of 287.70 eV appeared in C 1s of FSAH, belonging to the C=N bond in the hydrazone group, and the other three peaks BE shifted to 284.80, 286.28, and 288.70 eV, successively. The atom fraction at 284.80 eV increased to 70.47%, indicating that the dopamine grafting was successful. The peak of 535.90 eV in FSAH disappeared, the electron binding energy of –OH bond increased from 532.80 to 533.25 eV in O 1s, and the atomic fraction of the –OH decreased by 19.32% (from 55.96% to 36.64%). The N 1s curve formed by the Schiff base and aldimine condensation reactions, with BEs of

399.92, 400.15, and 400.92 eV for the C=N, N–N, and C–N bonds, respectively. The structure of the planned target adsorbent FSAH is compatible with these results.

2.1.3. TG and BET Analysis

As Figure 3a shows, the decomposition of SA and FSAH involved three processes: mass loss of 9% was attributed to the loss of water at around 210 °C in the first step; the thermal decomposition weight loss rate of carboxyl, hydrazone, amide, and hydroxyl groups increased clearly under 260 °C; and the breakdown and decomposition of the molecular carbon chain and basic skeleton (T > 260 °C).

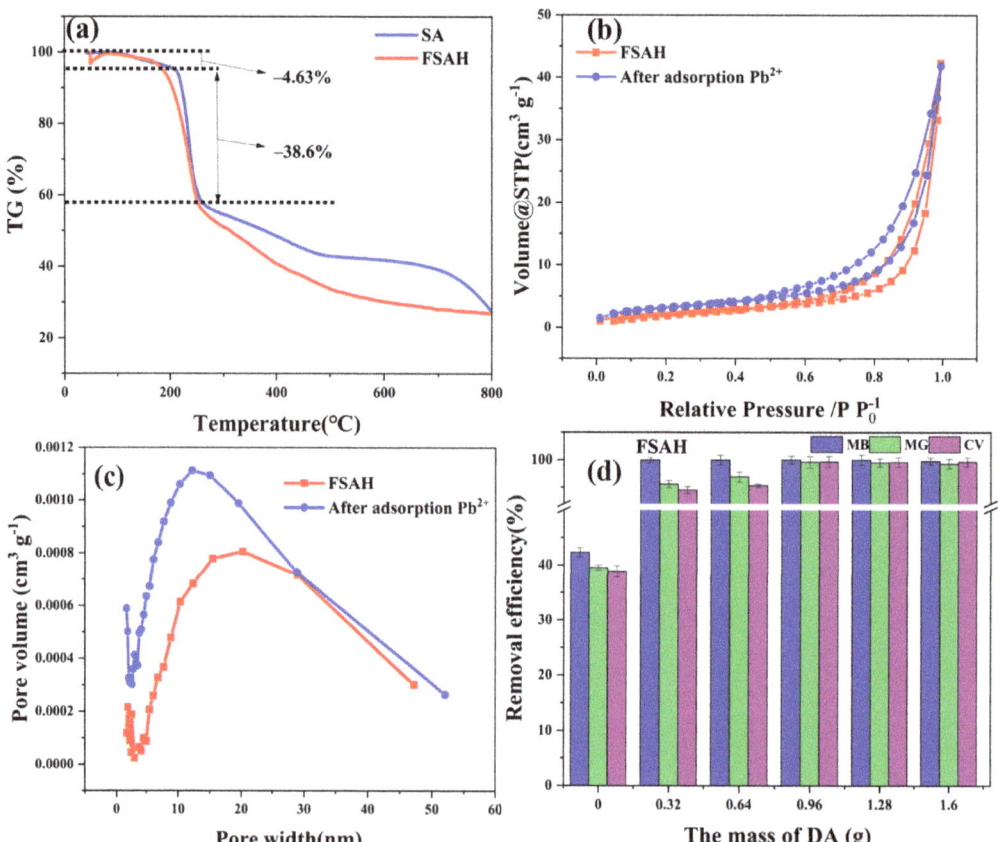

Figure 3. The TG curves of FSAH (**a**), SA BET N_2 adsorption and desorption isothermals curve for FSAH and FSAH after adsorption Pb^{2+} (**b**,**c**), the relationship between different DA mass ratios and dye removal rates (**d**).

Figure 3b,c showed that N_2 adsorption increases slowly under relatively low pressure, the adsorption capacity of N_2 increased obviously with the increase of pressure which can be classified as *VI* type adsorption isotherm according to IUPAC classification criteria, which showed a lag loop of H3 associated with slit pores when the relative pressure (P/P_0) is close to 0.6. After the adsorption of Pb^{2+} by FSAH, the average pore size of BJH decreased, and the concentrated distribution of pore size changed from 20.34 to 12.29 nm. The pore uniformity was good, but the volume of adsorption pores of BET and BJH increased greatly.

2.2. Adsorption Performance Study

2.2.1. DA Mass Ratio

The DA was applied to prepare FSAH to adsorb cationic dyes (MB, MG, and CV) with variable proportions (1.60, 1.28, 0.96, 0.64, and 0.32 g). When the amount of DA increased from 0 to 0.96 g, the removal rates of MB, MG, and CV increased from 42.31, 39.52, and 38.85 to 99.99, 99.61, and 99.66% (Figure 3d). The removal efficiency of cationic dyes did not change significantly when the amount of DA was increased. This indicated the carboxyl group on the SA chain was conjugated to a certain amount of DA by N-(3-dimethylaminopropyl)-N′-ethylcarbodiimide hydrochloride (EDC), and N-hydroxysuccinimide (NHS), and excessive DA remained in the PBS buffer. Considering cost and benefit, the input of DA grafting was determined to be 0.96 g.

2.2.2. pH

The relationship between pH and heavy metals (pH: 1–7) and dyes (pH: 1–12) removal rate is shown in Figure 4. The point of zero charge (pH$_{ZPC}$) value of FSAH at zero potential was 4.07. The amount of H$^+$ was increased, causing protonation of the adsorbent surface at pH < pH$_{ZPC}$, resulting in a positively charged surface. At pH > pH$_{ZPC}$, the level of OH$^-$ was increased, and functional groups of FSAH surfaces formed negative charge, thus promoting the removal rate of heavy metals and cationic dyes [26]. The pH of 5 was the best for FSAH to adsorb heavy metals and cationic dyes, because of the hydroxide formation at pH > 5; as Medusa software showed formation of copper hydroxide at pH = 6; and then the adsorption removal rate had no significant changes to dyes at pH > 5. When pH < pH$_{PZC}$, the adsorption ability was restricted because a mass of H$^+$ on FSAH surface caused group protonation, which prevented the adsorption removal rate [27,28].

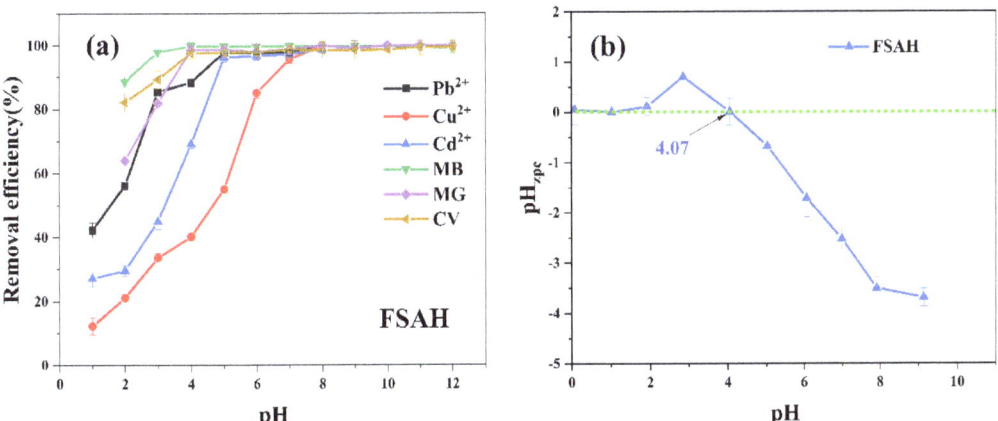

Figure 4. The relationship between pH and heavy metals and dyes removal rate (**a**); the pH$_{ZPC}$ of FSAH (**b**).

2.2.3. Adsorption Isotherm

To evaluate the equilibrium adsorption mechanism, the Langmuir [29], Freundlich, and Redlich–Peterson isotherm models [30] were used (Equations (S1)–(S4), respectively).

Compared with the fitting correlation coefficient (R^2) of the Freundlich isotherm model and Langmuir isotherm model, the Freundlich isotherm model was more suitable to describe the adsorption process of Pb^{2+}, Cd^{2+}, and Cu^{2+} by FSAH at three temperature conditions in Figure 5 ($R^2 > 0.95$). The adsorption intensity decreased with the increase in temperature ($1/n < 0.5$) and K_F increased with the increase of temperature, which indicated that the adsorption process was easier to proceed with accompanied by an increase in temperature. In the Langmuir isotherm model, Figure 5 shows the change curve of adsorption separation factor (R_L) and adsorbent initial concentration, $0 < R_L < 1$, indicating that the adsorption behavior is favorable and has a strong affinity (Figure S1). It can be inferred that the adsorption process of heavy metal ions mainly tended to be multilayer adsorption. In the adsorption process of MB, MG, and CV, the fitting correlation coefficient R^2 of the Langmuir isotherm model was >0.9, which was larger than the Freundlich isotherm model's R^2, and the adsorption data were more consistent with the Langmuir isotherm model. These results indicated that the adsorption of MB, MG, and CV by FSAH mainly involved monomolecular adsorption induced by the point-facing adsorption mechanism. K_F and n represented the adsorption strength and strength, respectively. The K_F and n values of MB were the largest when fitting parameters with the three cationic dyes, which indicated that the affinity of FSAH with cationic dyes was MB > CV > MG [31]. Rendering a high R^2 value and a low χ^2 value, fitting with the Redlich–Peterson model was also fairly good ($R^2 > 0.9$, Table S2). The saturated adsorption capacity of Pb^{2+}, Cd^{2+}, Cu^{2+}, MB, MG, and CV by FSAH were 371.4, 304.3, 157.1, 1147.71, 1332.75, and 1210.01 mg/g in Table S1, respectively. Comparison with other adsorption materials are reported in Table S6. The adsorption capacity of FSAH on heavy metals and dyes has a relatively large advantage. As a consequence, FSAH was a valuable adsorbent material for wastewater treatment.

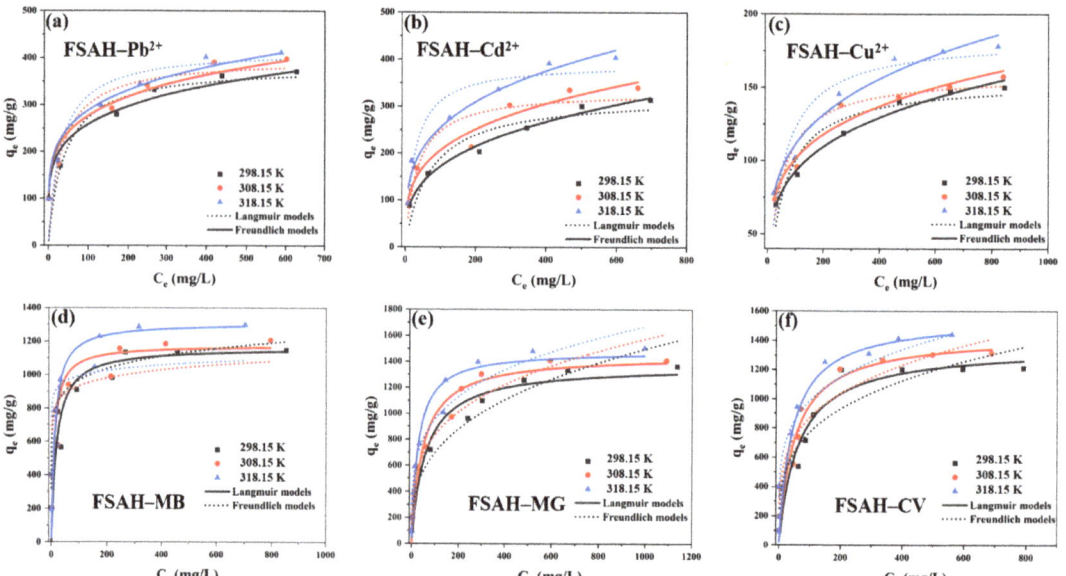

Figure 5. FSAH adsorption isotherm curves fitted by the Langmuir and Freundlich models to adsorb Pb^{2+} (**a**), Cd^{2+} (**b**), Cu^{2+} (**c**), MB (**d**), MG (**e**), and CV (**f**).

2.2.4. Adsorption Kinetic

In order to fit the experimental data, the non-linear pseudo-first-order (PFO) [32], pseudo-second-order (PSO) [33] rate laws kinetic models, and internal diffusion models are represented by Equations (S5)–(S7) in this paper, respectively.

Figure 6 and Tables S2 and S3 show that the adsorption of heavy metal ions and cationic dyes had a higher degree of fit with PSO, R^2(PSO) > R^2(PFO) and χ^2(PSO) < χ^2(PFO) based on high values of R^2 and low values of the non-linear chi-square statistics (χ^2), and the q_e calculated by PSO was more consistent with the actual value of the experiment. The relevant kinetic fitting parameters K_1, $K_2 > 0$, which showed that the adsorption behavior of heavy metals was spontaneous. In the case of the coexistence of chemical adsorption and physical adsorption, the rate-determining step was chemical adsorption. C_1, $C_2 \neq 0$ in the two stages of intraparticle diffusion adsorption, indicating that boundary layer diffusion has a great influence on adsorption. The $k_{id1} > k_{id2}$ which indicated that the surface of the adsorbent was occupied by the adsorbate with the increase of adsorption time, and the adsorbate molecules diffused from the surface of the adsorbent to the internal pores of the adsorbent, resulting in a very slow process and a decrease in the adsorption efficiency [34].

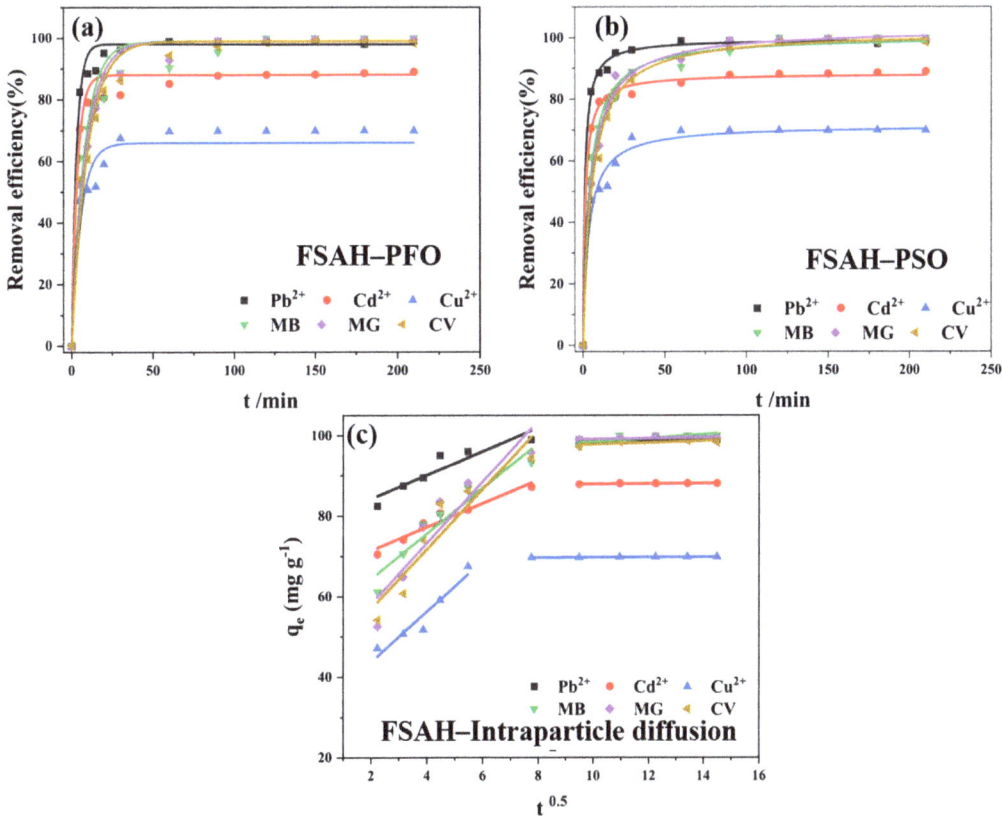

Figure 6. The PFO (**a**) and PSO kinetic plots (**b**), the intraparticle diffusion plots (**c**) for heavy metals and cationic dyes onto FSAH.

2.2.5. Thermodynamic Adsorption

The Gibbs free energy (ΔG^θ), enthalpy (ΔH^θ), and entropy (ΔS^θ) were calculated using Equations (S8)–(S11) at 298.15, 308.15, and 318.15 K, respectively [35,36]. Figure 7 and Table S4 show the related thermodynamic parameters. The $\Delta H > 0$ indicated that increasing the temperature increased the resultful collision efficiency between heavy metals, dyes, and FSAH, which promoted diffusion in the microchannels inside, improved adsorption ability to adsorb heavy metals which were very advantageous to adsorption. The $\Delta S > 0$ showed that solute adsorption was accompanied by solvent desorption in the adsorption process; the former process was accompanied by a loss in entropy, whereas the latter process increased entropy [37]. At various temperatures, the $\Delta G^\theta < 0$, adsorption processes were viable and spontaneous. In conclusion, adsorption reactions were spontaneous endothermic reactions.

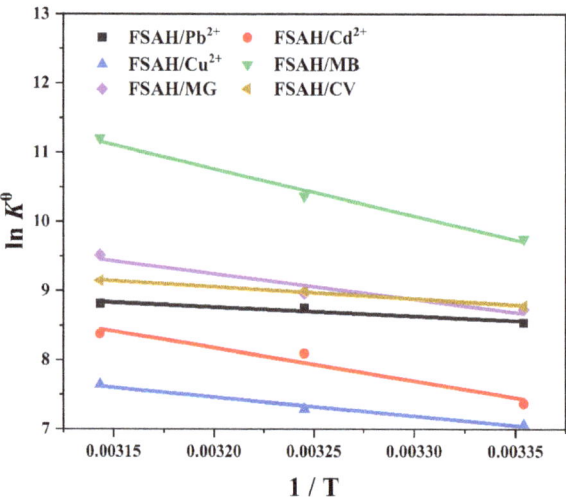

Figure 7. The adsorption capacity and the adsorption thermodynamics curve of heavy metal ions and dyes adsorption of FSAH at 298.15, 308.15, and 318.15 K, respectively.

2.2.6. Reuse Adsorption

To assess recycle adsorption ability of FSAH was an important index [38]. After five recycles, the adsorption removal rate of FSAH decreased to 93.15%, 77.28%, 55.18%, 92.45%, 89.23%, and 91.23% for Pb^{2+}, Cd^{2+}, Cu^{2+}, MB, MG, and CV (Figure 8), respectively. The main reasons were speculated on the partial functional groups that were occupied with heavy metals and dyes incomplete desorption, and the structure of FSAH could be damaged with HCl [39]. In general, the FSAH as heavy metals and dyes adsorbent materials possessed excellent regeneration and reuse potential.

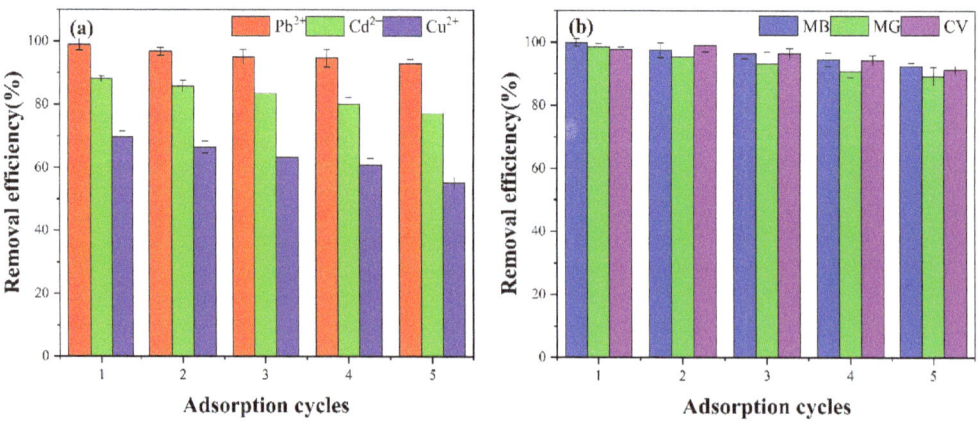

Figure 8. Removal efficiency of FSAH (a,b) adsorption of heavy metals and dyes after 5 recycles.

2.3. Adsorption Mechanism

Figure 9a shows the FT-IR analysis of FSAH before and after adsorption of Pb^{2+} and MB. The peak of C=N came from Schiff base condensation shifted from 1597 to 1586 cm^{-1}, The peak of COO$^-$ at 1392 cm^{-1} changed initial position, which showed that ion-exchange and coordination would have occurred after adsorption Pb^{2+}. The peak of carboxyl groups shifted significantly after adsorption MB, and new peaks appeared.

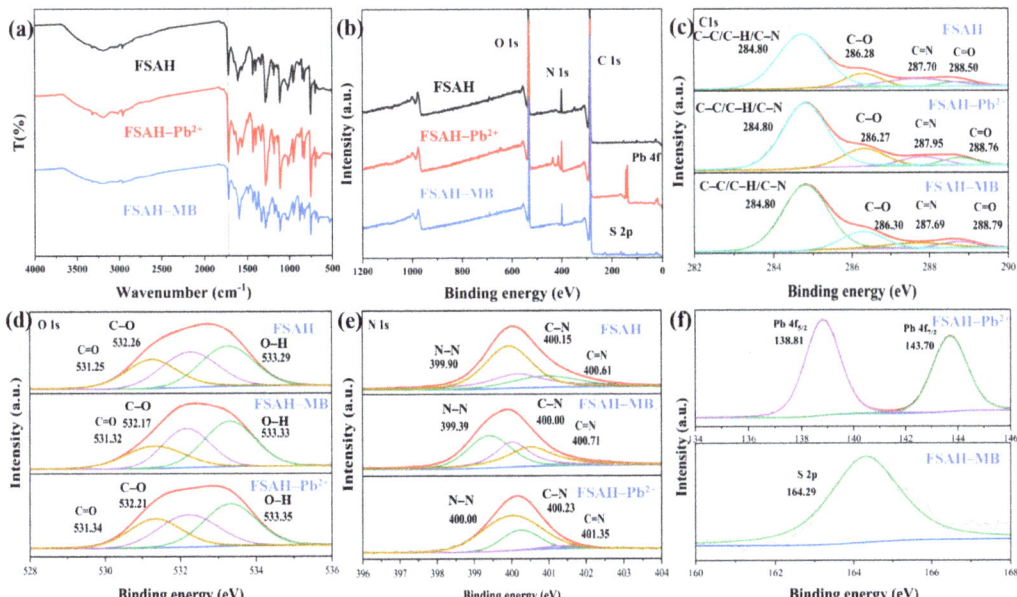

Figure 9. FT–IR (**a**) and XPS (**b**) analysis of the FSAH, and XPS analysis (**c,d**) of C 1s, O 1s, and N 1s (**e**) of the FSAH before and after adsorption of Pb^{2+} and MB, XPS analysis of Pb 4f and S 2p (**f**), respectively.

The new peaks of Pb 4f and S 2p clearly emerged in Figure 9b revealed that the FSAH had adsorbed Pb^{2+} and MB from XPS analysis, respectively. The O 1s peak for C–O of –COO⁻ shifted from 532.60 to 532.83 eV which possibly revealed the covalent interaction between Pb^{2+} and C–O of –COO⁻ was ionic interactions after adsorption Pb^{2+} [40–42]; however, the BE of C=N shifted from 401.28 to 402.05 eV (ΔBE > 0.5 eV), the atomic fractions of C=N were significantly reduced which indicated the preferential covalent interaction between Pb^{2+} and N-donor ligands over O-donors [10,43–45]. After adsorption of MB on FSAH, the S $2p_{3/2}$ (163.90 eV) and S $2p_{1/2}$ (168.60 eV) appeared in Figure 9c, and the O 1s spectra noticeably shifted, possibly from the influence of –OH and COO⁻ in FSAH. Because the dyes of MB, MG, and CV have positive-charge, FSAH has negative-charge when the system pH is greater than pH_{ZPC}, and the adsorption data of MB, MG, and CV fitted with the Langmuir model which showed that monolayer adsorption was dominant.

In general, the most possible explanation for adsorption is the ion exchange interaction from COO⁻ groups and chelation coordination from hydrazone groups played important roles in adsorption of heavy metals in FSAH; due to graft DA introducing functional groups that have more adsorption effect on dyes, especially catechol groups, which utilized electrostatic attraction, hydrogen bonding, π-π interaction, and Van der Waals force [46].

3. Conclusions

The secondary grafting of hydrazide and DA to prepare adsorption material FSAH, which has a high removal rate for Pb^{2+}, Cd^{2+}, Cu^{2+}, MB, MG, and CV. The saturated adsorption capacity of FSAH for Pb^{2+}, Cd^{2+}, Cu^{2+}, MB, MG, and CV were 371.4, 304.3, 157.1, 1147.71, 1332.75, and 1210.01 mg/g, respectively. SEM, BET, FT-IR, and XPS analysis show that heavy metals are easier to enter the FSAH in the spatial network structure; the structure of BDD boosted adsorption effectiveness of dyes through the π–π-interaction. At ideal pH = 5, the thermodynamic study revealed that the adsorption process is endothermic and spontaneous. The hydrazone groups coordination and ion exchange, which was primarily chemical adsorption, fitted the Freundlich model for heavy metal ion adsorption.

However, cationic dye adsorption was linked to a variety of interactions, including H-bond, electrostatic, and π-π interaction, which fitted the Langmuir model and revealed that monolayer adsorption was the most common. Furthermore, after fivefold adsorption–desorption, adsorption efficiency can still be over 80%. To sum up, we concluded that FSAH was a valuable adsorbent material for wastewater treatment.

4. Materials and Methods

4.1. Materials

SA (500–1000 mPa·s) was purchased from Adamas Reagent (Shanghai, China). Acetic acid, BDD, dopamine (DA), ethylene glycol, Pb $(NO_3)_2$, Cd $(NO_3)_2 \cdot 4H_2O$, Cu $(NO_3)_2 \cdot 4H_2O$, MB, MG, CV, EDC, NHS, lead, cadmium, and copper standard solution (1000 mg/L) were obtained from Aladdin Biochemical Technology (Shanghai, China). Anhydrous ethanol, and sodium periodate ($NaIO_4$) were obtained from the Kelong Chemical Reagent Factory (Chengdu, China). Unless otherwise noted, all reagents were used without further purification.

4.2. Preparation of DSA

SA (5 g) was dispersed in anhydrous ethanol (50 mL) for 2 h, and 40 mL aqueous solution (2.5 g, $NaIO_4$) was dropped into SA-ethanol dispersion in the dark for 12 h (318.15 K, pH = 4.0). Furthermore, ethylene glycol (5 mL) was used to terminate reaction, the product was washed 3 times with a mixture of ethanol and water (v/v = 5:4), and finally, DSA was vacuum-dried [47]. Using an automatic potentiometric titrator and the hydroxylamine hydrochloride method, the oxidation degree (OD) of DSA was determined by potentiometric titration for the detection of aldehyde groups [22,37], and the OD of DSA is 64.28%.

4.3. Preparation of FSAH

The mixture (DSA, 5.00 g and BDD, 2.70 g) was stirred to prepare DBD at 318 K for 12 h in ethanol (20 mL), which was collected by filtration; DBD (500 mg), EDC (440 mg), and NHS (295 mg) were stirred in PBS buffer solution (50 mL) for 30 min, and then added to 900 mg DA through nitrogen protection for 24 h at 15 °C. The hydrogel particle of FSAH was washed 3 times, and freeze-dried [18,21]. The preparation route of FSAH is presented in Figure 10.

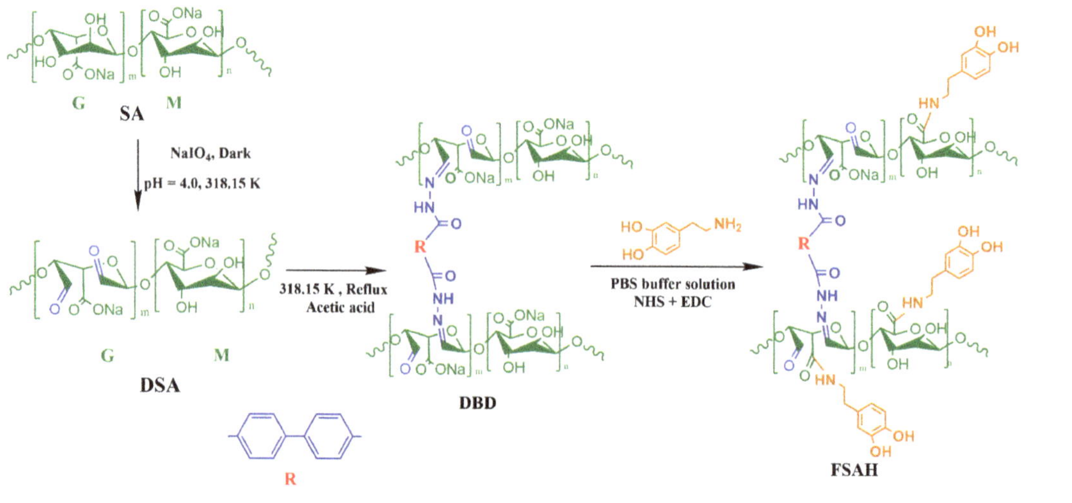

Figure 10. The preparation route of FSAH.

Supplementary Materials: The following supporting information can be downloaded at: https://www.mdpi.com/article/10.3390/gels8060343/s1, Figure S1: the change curve of adsorption separation factor R_L [29,30,48]. Table S1: Parameters calculated by Langmuir and Freundlich models for heavy metals and dyes adsorption onto FSAH (298.15 K, 308.1 5K, 318.15 K) [29,30]. Table S2. Kinetic parameters for heavy metals and cationic dyes adsorption onto FSAH [32,33,49]. Table S3. Intraparticle diffusion parameters for heavy metals and cationic dyes adsorption onto FSAH. Table S4. Thermodynamic parameters for the adsorption of heavy metal ions and dyes onto FSAH [35,36]. Figure S2. Effect of different salt ions on the adsorption capacity of FSAH for to adsorb Pb^{2+} (a), Cd^{2+} (b), Cu^{2+} (c), MB (d), MG (e), and CV (f) [50–54]. Table S5. EDS-mapping parameters of SA and FSAH, respectively. Table S6. Comparison with other adsorption materials [10,31,55–66].

Author Contributions: T.S.: Conceptualization, Methodology, Writing—original draft. Z.X.: Conceptualization, Supervision, X.M.: Funding acquisition, Investigation, Software. Y.F.: Investigation, Data curation. T.P.: Supervision, Resources. D.S.: Software. All authors have read and agreed to the published version of the manuscript.

Funding: This research was funded by A Project on Characteristic Key Laboratory of Guizhou Ordinary Colleges and Universities by Department of Education of Guizhou Province (Grant No: Qian Jiao He KY Zi [2018] 003) supported by Guizhou Provincial Science and Technology Projects. A project of Guizhou Province science and technology program (Grant No.: Qian Ke He Ji Chu [2019] 1295 supported by Guizhou Provincial Science and Technology Projects. Zunyi science and Technology Bureau of Guizhou Province and Moutai Institute Joint Science and Technology Cooperation Fund Project, Zunyi Ke He Hz [2021] No. 333.

Institutional Review Board Statement: Not applicable.

Informed Consent Statement: Not applicable.

Data Availability Statement: Not applicable.

Conflicts of Interest: The authors declare no conflict of interest.

References

1. Xie, R.; Jiang, W.; Wang, L.; Peng, J.; Chen, Y. Effect of pyrolusite loading on sewage sludge-based activated carbon in Cu(II), Pb(II), and Cd(II) adsorption. *Environ. Prog. Sustain. Energy* **2013**, *32*, 1066–1073. [CrossRef]
2. Shahzad, A.; Miran, W.; Rasool, K.; Nawaz, M.; Jang, J.; Lim, S.-R.; Lee, D.S. Heavy metals removal by EDTA-functionalized chitosan graphene oxide nanocomposites. *RSC Adv.* **2017**, *7*, 9764–9771. [CrossRef]
3. Akhavan, B.; Jarvis, K.; Majewski, P. Plasma polymer-functionalized silica particles for heavy metals removal. *ACS Appl. Mater. Interfaces* **2015**, *7*, 4265–4274. [CrossRef] [PubMed]
4. Singha, N.R.; Karmakar, M.; Mahapatra, M.; Mondal, H.; Dutta, A.; Deb, M.; Mitra, M.; Roy, C.; Chattopadhyay, P.K. An in situ approach for the synthesis of a gum ghatti-g-interpenetrating terpolymer network hydrogel for the high-performance adsorption mechanism evaluation of Cd(ii), Pb(ii), Bi(iii) and Sb(iii). *J. Mater. Chem. A* **2018**, *6*, 8078–8100. [CrossRef]
5. Singha, N.R.; Chattopadhyay, P.K.; Dutta, A.; Mahapatra, M.; Deb, M. Review on additives-based structure-property alterations in dyeing of collagenic matrices. *J. Mol. Liq.* **2019**, *293*, 111470. [CrossRef]
6. Shi, P.; Hu, X.; Duan, M. A UIO-66/tannic acid/chitosan/polyethersulfone hybrid membrane-like adsorbent for the dynamic removal of dye and Cr (VI) from water. *J. Clean. Prod.* **2021**, *290*, 125794. [CrossRef]
7. Cai, L.; Ying, D.; Liang, X.; Zhu, M.; Lin, X.; Xu, Q.; Cai, Z.; Xu, X.; Zhang, L. A novel cationic polyelectrolyte microsphere for ultrafast and ultra-efficient removal of heavy metal ions and dyes. *Chem. Eng. J.* **2021**, *410*, 128404. [CrossRef]
8. Saya, L.; Malik, V.; Singh, A.; Singh, S.; Gambhir, G.; Singh, W.R.; Chandra, R.; Hooda, S. Guar gum based nanocomposites: Role in water purification through efficient removal of dyes and metal ions. *Carbohydr. Polym.* **2021**, *261*, 117851. [CrossRef]
9. Khan, F.S.A.; Mubarak, N.M.; Tan, Y.H.; Khalid, M.; Karri, R.R.; Walvekar, R.; Abdullah, E.C.; Nizamuddin, S.; Mazari, S.A. A comprehensive review on magnetic carbon nanotubes and carbon nanotube-based buckypaper for removal of heavy metals and dyes. *J. Hazard. Mater.* **2021**, *413*, 125375. [CrossRef]
10. Usman, M.; Ahmed, A.; Yu, B.; Wang, S.; Shen, Y.; Cong, H. Simultaneous adsorption of heavy metals and organic dyes by beta-Cyclodextrin-Chitosan based cross-linked adsorbent. *Carbohydr. Polym.* **2021**, *255*, 117486. [CrossRef]
11. Chen, M.; Bi, R.; Zhang, R.; Yang, F.; Chen, F. Tunable surface charge and hydrophilicity of sodium polyacrylate intercalated layered double hydroxide for efficient removal of dyes and heavy metal ions. *Colloids Surf. A Physicochem. Eng. Asp.* **2021**, *617*, 126384. [CrossRef]
12. Simonescu, C.M.; Mason, T.J.; Calinescu, I.; Lavric, V.; Vinatoru, M.; Melinescu, A.; Culita, D.C. Ultrasound assisted preparation of calcium alginate beads to improve absorption of Pb(+2) from water. *Ultrason. Sonochem.* **2020**, *68*, 105191. [CrossRef]

13. Ding, H.; Zhang, X.; Yang, H.; Luo, X.; Lin, X. Highly efficient extraction of thorium from aqueous solution by fungal mycelium-based microspheres fabricated via immobilization. *Chem. Eng. J.* **2019**, *368*, 37–50. [CrossRef]
14. Gao, X.; Guo, C.; Hao, J.; Zhao, Z.; Long, H.; Li, M. Adsorption of heavy metal ions by sodium alginate based adsorbent—A review and new perspectives. *Int. J. Biol. Macromol.* **2020**, *164*, 4423–4434. [CrossRef]
15. Xue, S.; Xie, Z.; Chu, Y.; Yue, Y.; Shi, W.; Zhou, J. Synthesis of Sulfonylhydrazone Type Probe with High Selectivity for Rapid Detection of Mercury and Its Application in Adsorption and HeLa Cell. *Chin. J. Org. Chem.* **2021**, *41*, 1138–1145. [CrossRef]
16. Xue, S.; Xie, Z.; He, J. 1-[(2-Hydroxy-phenylimino)-methyl]-naphthalen-2-ol: Application in detection and adsorption of aluminum ions. *Res. Chem. Intermed.* **2021**, *47*, 4333–4347. [CrossRef]
17. Wang, H.; Xue, S.; Zhou, X.; Liu, J.; Xie, Z. Synthesis of highly selective copper ion probe and its application in adsorption. *Chin. J. Lumin.* **2021**, *42*, 1427. [CrossRef]
18. Ma, J.; Fang, S.; Shi, P.; Duan, M. Hydrazine-Functionalized guar-gum material capable of capturing heavy metal ions. *Carbohydr. Polym.* **2019**, *223*, 115137. [CrossRef]
19. Wu, C.; Wang, H.; Wei, Z.; Li, C.; Luo, Z. Polydopamine-mediated surface functionalization of electrospun nanofibrous membranes: Preparation, characterization and their adsorption properties towards heavy metal ions. *Appl. Surf. Sci.* **2015**, *346*, 207–215. [CrossRef]
20. Ye, Q.; Zhou, F.; Liu, W. Bioinspired catecholic chemistry for surface modification. *Chem. Soc. Rev.* **2011**, *40*, 4244–4258. [CrossRef]
21. Chen, T.; Chen, Y.; Rehman, H.U.; Chen, Z.; Yang, Z.; Wang, M.; Li, H.; Liu, H. Ultratough, Self-Healing, and Tissue-Adhesive Hydrogel for Wound Dressing. *ACS Appl. Mater. Interfaces* **2018**, *10*, 33523–33531. [CrossRef]
22. Wang, L.; Hou, Y.; Zhong, X.; Hu, J.; Shi, F.; Mi, H. Preparation and catalytic performance of alginate-based Schiff Base. *Carbohydr. Polym.* **2019**, *208*, 42–49. [CrossRef]
23. Wilson, L.D.; Pratt, D.Y.; Kozinski, J.A. Preparation and sorption studies of beta-cyclodextrin-chitosan-glutaraldehyde terpolymers. *J. Colloid Interface Sci.* **2013**, *393*, 271–277. [CrossRef]
24. Gorzalski, A.S.; Donley, C.; Coronell, O. Elemental composition of membrane foulant layers using EDS, XPS, and RBS. *J. Membr. Sci.* **2017**, *522*, 31–44. [CrossRef]
25. He, Y.; Gou, S.; Zhou, L.; Tang, L.; Liu, T.; Liu, L.; Duan, M. Amidoxime-functionalized polyacrylamide-modified chitosan containing imidazoline groups for effective removal of Cu^{2+} and Ni^{2+}. *Carbohydr. Polym.* **2021**, *252*, 117160. [CrossRef]
26. Pashaei-Fakhri, S.; Peighambardoust, S.J.; Foroutan, R.; Arsalani, N.; Ramavandi, B. Crystal violet dye sorption over acrylamide/graphene oxide bonded sodium alginate nanocomposite hydrogel. *Chemosphere* **2021**, *270*, 129419. [CrossRef]
27. Li, Y.; Zhou, Y.; Zhou, Y.; Lei, J.; Pu, S. Cyclodextrin modified filter paper for removal of cationic dyes/Cu ions from aqueous solutions. *Water Sci. Technol.* **2018**, *78*, 2553–2563. [CrossRef]
28. Huang, Q.; Liu, M.; Zhao, J.; Chen, J.; Zeng, G.; Huang, H.; Tian, J.; Wen, Y.; Zhang, X.; Wei, Y. Facile preparation of polyethylenimine-tannins coated SiO_2 hybrid materials for Cu^{2+} removal. *Appl. Surf. Sci.* **2018**, *427*, 535–544. [CrossRef]
29. Langmuir, I. The Constitution and Fundamental Properties of Solids and Liquids. Part I. Solids. *J. Am. Chem. Soc.* **1916**, *38*, 2221–2295. [CrossRef]
30. Freundlich, H. Über die Adsorption in Lösungen. *Z. Phys. Chem.* **1906**, *57U*, 385–470. [CrossRef]
31. Lin, L.; Tang, S.; Wang, X.; Sun, X.; Yu, A. Hexabromocyclododecane alters malachite green and lead(II) adsorption behaviors onto polystyrene microplastics: Interaction mechanism and competitive effect. *Chemosphere* **2021**, *265*, 129079. [CrossRef] [PubMed]
32. Lagergren, S. Zur theorie der sogenannten adsorption geloster stoffe. *K. Sven. Vetensk. Handl.* **1898**, *24*, 1–39.
33. Ho, Y.S.; McKay, G. Sorption of dye from aqueous solution by peat. *Chem. Eng. J.* **1998**, *70*, 115–124. [CrossRef]
34. Cui, L.; Wang, Y.; Hu, L.; Gao, L.; Du, B.; Wei, Q. Mechanism of Pb(ii) and methylene blue adsorption onto magnetic carbonate hydroxyapatite/graphene oxide. *RSC Adv.* **2015**, *5*, 9759–9770. [CrossRef]
35. Cestari, A.R.; Vieira, E.F.; Tavares, A.M.; Bruns, R.E. The removal of the indigo carmine dye from aqueous solutions using cross-linked chitosan: Evaluation of adsorption thermodynamics using a full factorial design. *J. Hazard. Mater.* **2008**, *153*, 566–574. [CrossRef]
36. Zhang, N.; Zhang, H.; Li, R.; Xing, Y. Preparation and adsorption properties of citrate-crosslinked chitosan salt microspheres by microwave assisted method. *Int. J. Biol. Macromol.* **2020**, *152*, 1146–1156. [CrossRef]
37. Shi, T.; Xie, Z.; Zhu, Z.; Shi, W.; Liu, Y.; Liu, M.; Mo, X. Effective removal of metal ions and cationic dyes from aqueous solution using different hydrazine-dopamine modified sodium alginate. *Int. J. Biol. Macromol.* **2022**, *195*, 317–328. [CrossRef]
38. Gu, P.; Zhang, S.; Li, X.; Wang, X.; Wen, T.; Jehan, R.; Alsaedi, A.; Hayat, T.; Wang, X. Recent advances in layered double hydroxide-based nanomaterials for the removal of radionuclides from aqueous solution. *Environ. Pollut.* **2018**, *240*, 493–505. [CrossRef]
39. Zheng, L.; Zhang, S.; Cheng, W.; Zhang, L.; Meng, P.; Zhang, T.; Yu, H.; Peng, D. Theoretical calculations, molecular dynamics simulations and experimental investigation of the adsorption of cadmium(ii) on amidoxime-chelating cellulose. *J. Mater. Chem. A* **2019**, *7*, 13714–13726. [CrossRef]
40. Dai, J.; Yan, H.; Yang, H.; Cheng, R. Simple method for preparation of chitosan/poly(acrylic acid) blending hydrogel beads and adsorption of copper(II) from aqueous solutions. *Chem. Eng. J.* **2010**, *165*, 240–249. [CrossRef]
41. Dong, Y.; Ma, Y.; Zhai, T.; Shen, F.; Zeng, Y.; Fu, H.; Yao, J. Silver Nanoparticles Stabilized by Thermoresponsive Microgel Particles: Synthesis and Evidence of an Electron Donor-Acceptor Effect. *Macromol. Rapid Commun.* **2007**, *28*, 2339–2345. [CrossRef]

42. Ling, C.; Liu, F.-Q.; Long, C.; Chen, T.-P.; Wu, Q.-Y.; Li, A.-M. Synergic removal and sequential recovery of acid black 1 and copper (II) with hyper-crosslinked resin and inside mechanisms. *Chem. Eng. J.* **2014**, *236*, 323–331. [CrossRef]
43. Fang, Y.; Luo, B.; Jia, Y.; Li, X.; Wang, B.; Song, Q.; Kang, F.; Zhi, L. Renewing Functionalized Graphene as Electrodes for High-Performance Supercapacitors. *Adv. Mater.* **2012**, *24*, 6348–6355. [CrossRef]
44. Wang, Y.-H.; Bayatpour, S.; Qian, X.; Frigo-Vaz, B.; Wang, P. Activated carbon fibers via reductive carbonization of cellulosic biomass for adsorption of nonpolar volatile organic compounds. *Colloids Surf. A Physicochem. Eng. Asp.* **2021**, *612*, 125908. [CrossRef]
45. He, S.; Zhang, F.; Cheng, S.; Wang, W. Synthesis of Sodium Acrylate and Acrylamide Copolymer/GO Hydrogels and Their Effective Adsorption for Pb^{2+} and Cd^{2+}. *ACS Sustain. Chem. Eng.* **2016**, *4*, 3948–3959. [CrossRef]
46. Liu, M.; Xie, Z.; Ye, H.; Li, W.; Shi, W.; Liu, Y.; Zhang, Y. Waste polystyrene foam—Chitosan composite materials as high-efficient scavenger for the anionic dyes. *Colloids Surf. A Physicochem. Eng. Asp.* **2021**, *627*, 127155. [CrossRef]
47. Wu, L.; Li, L.; Pan, L.; Wang, H.; Bin, Y. MWCNTs reinforced conductive, self-healing polyvinyl alcohol/carboxymethyl chitosan/oxidized sodium alginate hydrogel as the strain sensor. *J. Appl. Polym. Sci.* **2020**, *138*, 49800. [CrossRef]
48. Milani, S.A.; Karimi, M. Isotherm, kinetic and thermodynamic studies for Th(IV) sorption by amino group-functionalized titanosilicate from aqueous solutions. *Korean J. Chem. Eng.* **2017**, *34*, 1159–1169. [CrossRef]
49. Chen, X.; Li, P.; Zeng, X.; Kang, Y.; Wang, J.; Xie, H.; Zhang, Y. Efficient adsorption of methylene blue by xanthan gum derivative modified hydroxyapatite. *Int. J. Biol. Macromol.* **2020**, *151*, 1040–1048. [CrossRef]
50. Dong, J.; Du, Y.; Duyu, R.; Shang, Y.; Zhang, S.; Han, R. Adsorption of copper ion from solution by polyethylenimine modified wheat straw. *Bioresour. Technol. Rep.* **2019**, *6*, 96–102. [CrossRef]
51. Wang, J.; Xu, L.; Cheng, C.; Meng, Y.; Li, A. Preparation of new chelating fiber with waste PET as adsorbent for fast removal of Cu^{2+} and Ni^{2+} from water: Kinetic and equilibrium adsorption studies. *Chem. Eng. J.* **2012**, *193–194*, 31–38. [CrossRef]
52. Yadav, S.; Asthana, A.; Singh, A.K.; Chakraborty, R.; Vidya, S.S.; Susan, M.; Carabineiro, S.A.C. Adsorption of cationic dyes, drugs and metal from aqueous solutions using a polymer composite of magnetic/beta-cyclodextrin/activated charcoal/Na alginate: Isotherm, kinetics and regeneration studies. *J. Hazard. Mater.* **2021**, *409*, 124840. [CrossRef]
53. Chen, H.; Gao, B.; Li, H. Removal of sulfamethoxazole and ciprofloxacin from aqueous solutions by graphene oxide. *J. Hazard. Mater.* **2015**, *282*, 201–207. [CrossRef]
54. Chen, Y.; Lan, T.; Duan, L.; Wang, F.; Zhao, B.; Zhang, S.; Wei, W. Adsorptive Removal and Adsorption Kinetics of Fluoroquinolone by Nano-Hydroxyapatite. *PLoS ONE* **2015**, *10*, e0145025. [CrossRef]
55. Chen, Y.; Wang, S.; Li, Y.; Liu, Y.; Chen, Y.; Wu, Y.; Zhang, J.; Li, H.; Peng, Z.; Xu, R.; et al. Adsorption of Pb(II) by tourmaline-montmorillonite composite in aqueous phase. *J. Colloid Interface Sci.* **2020**, *575*, 367–376. [CrossRef]
56. Hu, R.; Wang, X.; Dai, S.; Shao, D.; Hayat, T.; Alsaedi, A. Application of graphitic carbon nitride for the removal of Pb(II) and aniline from aqueous solutions. *Chem. Eng. J.* **2015**, *260*, 469–477. [CrossRef]
57. Alqadami, A.A.; Naushad, M.; ZA, A.L.; Alsuhybani, M.; Algamdi, M. Excellent adsorptive performance of a new nanocomposite for removal of toxic Pb(II) from aqueous environment: Adsorption mechanism and modeling analysis. *J. Hazard. Mater.* **2020**, *389*, 121896. [CrossRef]
58. Thanarasu, A.; Periyasamy, K.; Manickam Periyaraman, P.; Devaraj, T.; Velayutham, K.; Subramanian, S. Comparative studies on adsorption of dye and heavy metal ions from effluents using eco-friendly adsorbent. *Mater. Today Proc.* **2021**, *36*, 775–781. [CrossRef]
59. Sheikhi, A.; Safari, S.; Yang, H.; van de Ven, T.G.M. Copper Removal Using Electrosterically Stabilized Nanocrystalline Cellulose. *ACS Appl. Mater. Interfaces* **2015**, *7*, 11301–11308. [CrossRef]
60. Huang, Y.; Wang, Z. Preparation of composite aerogels based on sodium alginate, and its application in removal of Pb^{2+} and Cu^{2+} from water. *Int. J. Biol. Macromol.* **2018**, *107*, 741–747. [CrossRef]
61. Ilgin, P.; Durak, H.; Gür, A. A Novel pH-Responsive p(AAm-co-METAC)/MMT Composite Hydrogel: Synthesis, Characterization and Its Absorption Performance on Heavy Metal İons. *Polym.-Plast. Technol. Eng.* **2015**, *54*, 603–615. [CrossRef]
62. Ali, E.A.M.; Sayed, M.A.; Abdel-Rahman, T.M.A.; Hussein, R. Fungal remediation of Cd(ii) from wastewater using immobilization techniques. *RSC Adv.* **2021**, *11*, 4853–4863. [CrossRef] [PubMed]
63. Saeed, A.; Sharif, M.; Iqbal, M. Application potential of grapefruit peel as dye sorbent: Kinetics, equilibrium and mechanism of crystal violet adsorption. *J. Hazard. Mater.* **2010**, *179*, 564–572. [CrossRef] [PubMed]
64. Zhang, S.; Zhang, F.; Yang, M.; Fang, P. POSS modified Ni_xO_y-decorated TiO_2 nanosheets: Nanocomposites for adsorption and photocatalysis. *Appl. Surf. Sci.* **2021**, *566*, 150604. [CrossRef]
65. Ji, Q.; Li, H. High surface area activated carbon derived from chitin for efficient adsorption of Crystal Violet. *Diam. Relat. Mater.* **2021**, *118*, 108516. [CrossRef]
66. Abebe, M.W.; Kim, H. Methylcellulose/tannic acid complex particles coated on alginate hydrogel scaffold via Pickering for removal of methylene blue from aqueous and quinoline from non-aqueous media. *Chemosphere* **2021**, *286*, 131597. [CrossRef]

MDPI
St. Alban-Anlage 66
4052 Basel
Switzerland
www.mdpi.com

Gels Editorial Office
E-mail: gels@mdpi.com
www.mdpi.com/journal/gels

Disclaimer/Publisher's Note: The statements, opinions and data contained in all publications are solely those of the individual author(s) and contributor(s) and not of MDPI and/or the editor(s). MDPI and/or the editor(s) disclaim responsibility for any injury to people or property resulting from any ideas, methods, instructions or products referred to in the content.

www.ingramcontent.com/pod-product-compliance
Lightning Source LLC
LaVergne TN
LVHW070606100526
838202LV00012B/574